United States Nuclear Regulatory Commission

Protecting People and the Environment

NUREG-2117, Rev. 1

I0482762

Practical Implementation Guidelines for SSHAC Level 3 and 4 Hazard Studies

Office of Nuclear Regulatory Research

AVAILABILITY OF REFERENCE MATERIALS
IN NRC PUBLICATIONS

United States Nuclear Regulatory Commission

Protecting People and the Environment

NUREG-2117, Rev. 1

Practical Implementation Guidelines for SSHAC Level 3 and 4 Hazard Studies

Manuscript Completed: May 2011
Date Published: April 2012

Prepared by:
Annie M. Kammerer
Jon P. Ake

NRC Project Manager:
Richard Rivera-Lugo

Office of Nuclear Regulatory Research

ABSTRACT

10 CFR 100.23, paragraphs (c) and (d) require that the geological, seismological, and engineering characteristics of a site and its environs be investigated in sufficient scope and detail to permit an adequate evaluation of the Safe Shutdown Earthquake (SSE) Ground Motion for the site. In addition, 10 CFR 100.23, paragraph (d)(1), *"Determination of the Safe Shutdown Earthquake Ground Motion,"* requires that uncertainty inherent in estimates of the SSE be addressed through an appropriate analysis such as a probabilistic seismic hazard analysis (PSHA). In response to these requirements, in 1997, the U.S. Nuclear Regulatory Commission published NUREG/CR-6372, *Recommendations for Probabilistic Seismic Hazard Analysis. Guidance on Uncertainty and the Use of Experts.* Written by the Senior Seismic Hazard Analysis Committee (SSHAC), NUREG/CR-6372 provides guidance regarding the manner in which the uncertainties in PSHA should be addressed using expert judgment. In the 15 years since its publication, NUREG/CR-6372 has provided many PSHA studies with the framework and guidance that have come to be known simply as the "SSHAC Guidelines." The information in this NUREG is based on recent efforts to capture the lessons learned in the PSHA studies that have been undertaken using the SSHAC Guidelines. As a companion to NUREG/CR-6372, this NUREG provides additional practical implementation guidelines consistent with the framework and higher-level guidance of the SSHAC Guidelines.

FOREWORD

The complexity of tectonic environments and the limited data available for seismic source and ground motion characterization make the use of a significant level of expert judgment in seismic hazard assessment studies unavoidable. In the mid-90s the Nuclear Regulatory Commission, the U.S. Department of Energy, and the Electric Power Research Institute sponsored a study to develop recommendations for how studies incorporating the use of expert assessments should be conducted in the future. The Senior Seismic Hazard Analysis Committee (SSHAC) developed a structured, multi-level assessment process (the "SSHAC process") described in NUREG/CR-6372 that has since been used for numerous natural hazard studies and is recommended in NRC Regulatory Guide 1.208 for the development of new models to be used in probabilistic seismic hazard analyses.

This NUREG represents a companion document to NUREG/CR-6372 and is expected to be used for seismic hazard assessment whenever NRC guidance or communications call for use of NUREG/CR-6372 or the SSHAC guidelines. This NUREG was developed after careful review and assessment of lessons learned during the many projects that have been undertaken using the SSHAC guidelines since their publication in 1997.

The objectives of the additional practical guidance provided in this NUREG are: (1) determination of more accurate and consistent assessments of seismic hazard and the associated uncertainty, (2) standardization and complete and transparent documentation of the assessment process undertaken, the input data, and the basis for the resulting model and findings, (3) increased regulatory assurance based on the transparency of the study's technical basis and, (4) the increased longevity of a study as a result of the ability to assess new data against the existing model and its basis and assumptions. All of these goals lead to greater regulatory assurance and stability.

The guidance in this document applies equally to hazard assessment studies performed for design of new nuclear facilities and reassessment of operating reactors.

TABLE OF CONTENTS

LIST OF FIGURES

LIST OF TABLES

EXECUTIVE SUMMARY

In 1997, the U.S. Nuclear Regulatory Commission (NRC) issued NUREG/CR-6372 entitled, *Recommendations for Probabilistic Seismic Hazard Analysis: Guidance on Uncertainty and the Use of Experts*. The document was the culmination of 4 years of deliberations by the Senior Seismic Hazard Analysis Committee (SSHAC) regarding the manner in which the uncertainties in probabilistic seismic hazard analysis (PSHA) should be addressed using expert judgment. The document describes a formal process for structuring and conducting expert assessments that has come to be known as a "SSHAC process," and the recommendations made in the report are referred to as the SSHAC guidelines.

The SSHAC guidelines defined four levels at which hazard assessment studies can be conducted, ranging from the simplest (Level 1) to the most complicated and demanding (Level 4). The SSHAC report focused a great deal of attention on the conduct of Level 4 studies but provided comparatively little guidance on the lower levels of study, particularly Level 3.

This NUREG serves two primary purposes—it provides (1) additional levels of detail on topics related to the implementation of SSHAC processes beyond those provided in the original SSHAC report, particularly for Level 3 studies, and (2) additional guidance on the implementation of Level 3 and 4 studies in light of experience gained from past SSHAC projects. Over the past 15 years, several SSHAC Level 3 and 4 studies have been conducted, thus leading to an expanded "database" of experience in the intricacies of carrying out the SSHAC process in actual projects.

This document is intended to complement—rather than replace—the SSHAC guidelines report. The recommendations given in the SSHAC document regarding methodology were made largely in the abstract without the benefit of significant experience in many areas that would allow for specific guidance. As a result, they provide a very useful framework but are generally at a very high level. This document is intended to fill in that framework with details (necessarily) missing in the original SSHAC guidelines. In addition, the SSHAC guidelines devoted most effort to discussing SSHAC Level 4 studies and provided very little guidance regarding the specific approaches that would be appropriate for Level 3 studies. For this reason, this document—and the recent work supporting it—focus a significant amount of effort on developing and describing the appropriate methods and approaches for a Level 3 study.

This report summarizes the significant studies and case histories that led to the development of SSHAC guidance and then further summarizes the SSHAC studies that have occurred up to the present time. Several notable multi-expert probabilistic hazard studies and guidance studies conducted over the past 3 decades have been reviewed with the intent of showing the evolution of studies that led to the development of the SSHAC report in 1997 and the studies that have followed. The post-1997 studies (and those that occurred during the 4 years that SSHAC was being developed) include only those studies that followed SSHAC Level 3 or 4 processes.

Because of the need to use expert judgment, the probabilistic hazard community explored the decision analysis field of "expert elicitation" early on. Indeed, the terminology of "expert elicitation" has permeated the probabilistic seismic hazard literature for many years, and the SSHAC report refers to the process as a 'formal, structured expert elicitation process." However, the development of the SSHAC guidance represented a departure from classic expert elicitation processes. Now, with considerable experience from multiple projects, the SSHAC process has reached a point in its evolution that a distinction is required with classic expert elicitation. It is more appropriate to refer to the SSHAC process as multiple-expert *assessment* because it differs from classical expert *elicitation* in a number of ways. One key difference is the nature of the output required from the experts. As a result, the definition of "expert" also differs because there is at least as much, if not more, emphasis on subject expertise rather than normative expertise. Another important difference is the expectation that, in SSHAC-based studies, experts interact rather than remain entirely independent from one another. The other respect in which the SSHAC process is distinct from classical expert elicitation is that the assessments of individual experts are integrated rather than aggregated. The following paragraphs discuss each of these differences in more detail.

A review of the history of expert assessment studies and the development of associated guidance documents (as described above) indicates the following points: (1) the existing SSHAC guidance effectively defined basic concepts that have proven to be useful in actual application, (2) sufficient actual project experience with SSHAC Level 3 and 4 studies exists to be able to benefit from lessons learned, and (3) a need exists to develop new implementation guidance that details what works well and helps the reader avoid the pitfalls. Expert assessment methodologies generally, and SSHAC implementation processes specifically, are evolutionary in nature and continually improve with each project application.

The following key statement in the SSHAC guidelines encapsulates the ethos of the SSHAC approach: *"Regardless of the scale of the PSHA study the goal remains the same: to represent the center, the body, and the range that the larger informed technical community would have if they were to conduct the study"* (NUREG/CR-6372). For brevity, the "center, body, and range of the informed technical community" is indicated as "CBR of the ITC." A key word in the concept is "informed," which the SSHAC guidelines specifically define as an expert who has full access to the complete database developed for a project and has fully participated in the interactive SSHAC process. In other words, the selected experts who participate in the PSHA study must endeavor to represent *"the larger informed technical community"* by assuming the hypothetical case where the others in the larger technical community become "informed" through participation in the same process. The SSHAC guidelines recognize that this is a hypothetical exercise, but the goal would be to ensure that a broad range of views is considered. In practice, however, the term "informed" is often either ignored or misinterpreted as simply meaning "expert in the field of interest." Thus, some view the process of capturing or representing the CBR or the ITC as a process of somehow conducting a poll or surveying the larger community for its opinions.

In the spirit of maintaining the fundamental SSHAC objective and clarifying the concept with terms that reflect actual practice, this report presents an alternative statement of the fundamental objective of the SSHAC process. This alternate description explains that the objective of the SSHAC guidance is actually achieved through a two-stage process of *evaluation* followed by *integration*. Therefore, consistent with the original intent of the SSHAC guidance, we recast the goals of the SSHAC process in terms of the two main activities (i.e., evaluation and integration) by the following statement:

> The fundamental goal of a SSHAC process is to carry out properly and document completely the activities of evaluation and integration, defined as:
>
> *Evaluation:* The consideration of the complete set of data, models, and methods proposed by the larger technical community that are relevant to the hazard analysis.

Integration: Representing the center, body, and range of technically defensible interpretations in light of the evaluation process (i.e., informed by the assessment of existing data, models, and methods).

In light of these definitions, we propose that it is clearer to refer to the CBR of the "technically defensible interpretations" (TDI) instead of CBR of the ITC. However, it is important to emphasize that the careful evaluation of the larger technical community's viewpoints remains a vital part of the SSHAC process. We simply have removed the term "informed" because of its specialized definition in the original SSHAC guidelines. Similarly, we propose to replace the term "community distribution" that is used frequently in the original SSHAC guidelines to describe the outcome from a SSHAC assessment process with the term "integrated distribution." This change of terms will remove any perception that we arrived at the final assessments and models through a mere poll of the community.

The original SSHAC guidelines stated that "*it is absolutely necessary that there be a clear definition of ownership of the inputs into the PSHA (and hence ownership of the results of the PSHA)*" (NUREG/CR-6372). Regardless of the SSHAC Level (termed Study Level in the report) of the study, the definition of ownership is indeed extremely important both because it focuses the attention of the project participants on the importance of their individual contribution to the overall product and because it is key to successfully involving multiple experts in the overall process. The term "ownership" in this context is different from the sense of property ownership whereby the project sponsors have legal ownership of the project deliverables. Instead, the SSHAC guidelines focus on the concept of intellectual ownership of the results, which means taking responsibility for the robustness and defensibility of the various inputs to the hazard calculations. Intellectual ownership of these inputs therefore implies being able and willing to provide full and detailed explanation of the technical bases for all the decisions that led to the models and parameter values entered into the hazard calculations as well as for the rationale behind the relative weighting of these choices on the branches of the logic-tree.

The original SSHAC guidelines conveyed the impression that the biggest step from one SSHAC Level to another is between Levels 3 and 4 because Levels 1, 2, and 3 are all based on the concept of a Technical Integrator (TI) whereas a Level 4 study introduces the concept of the Technical Facilitator Integrator (TFI). In practice, the important differentiation in terms of complexity, cost, and schedule is between the simpler processes of Levels 1 and 2 and the more involved processes of Levels 3 and 4, which both include workshops, the participation of a participatory peer review panel, and generally larger groups of evaluators.

Because adopting a Level 3 or a Level 4 process to conduct a PSHA results in a significant increase in the cost and duration of the study over that required to conduct a Level 1 or Level 2 project, it is important to highlight the potential benefits to be gained by moving to these higher levels. These benefits are associated with the greater levels of regulatory assurance in Level 3 and 4 studies. We define *regulatory assurance* to mean confidence on the part of the NRC (or other regulator or reviewer) that the data, models, and methods of the larger technical community have been properly considered and that the center, body, and range of technically defensible interpretations have been appropriately represented and documented. In other words, it is increased confidence that the basic objectives of a SSHAC process have been met. We do not use the term "reasonable assurance" because it has a specific definition within the NRC's regulatory framework related to compliance with regulations. Rather, regulatory assurance is a qualitative term that is specific to the confidence that is engendered by the proper execution of a SSHAC process.

A stable assessment of the hazard at a site, conducted in a transparent process with the participation of several experts under the continuous review of a panel of experienced experts, provides greater assurance to a regulator that uncertainties have been effectively captured. In turn, a strong hazard study provides the underpinnings of the design basis ground motions for a critical facility, and the technical basis for the PSHA can be easily defended should contentions be raised. This increased regulatory assurance is the primary benefit obtained by conducting a Level 3 or Level 4 study. However, adoption of a Level 3 or 4 process does not guarantee regulatory acceptance even if the project fully conforms to the procedural requirements.

Data collection and processing should be considered principal activities in seismic hazard studies (particularly for studies conducted for important sites) because of their potential to reduce key uncertainties in the hazard inputs. An advantage of a well-structured SSHAC process is that it can be used to identify specific data collection activities that have highest potential to reduce the most hazard-significant uncertainties. A trade-off always exists between the resources required to conduct new data collection activities and the potential for uncertainty reduction. There will be situations where pressures of budget and schedule preclude extensive acquisition or collection of new data, although attention should be paid to the data collection requirements specified in regulatory guidance such as RG 1.208 (USNRC, 2007). However, even in cases where budget and schedule constraints preclude collection of new data, comprehensive data compilation (gathering and ordering all existing information) is an indispensable core requirement. Expert judgment should only be used to identify and quantify the uncertainty that remains after appropriate data collection and analysis activities have been completed.

SSHAC Level 3 and 4 processes provide a structured and transparent framework for conducting multiple-expert assessments that effectively capture epistemic uncertainty in hazard analyses. Central to the success of the process is the clear definition of the different roles that experts play and how they interact in the process. It is important to understand the role of the individual or group in the process, the responsibilities that the individual or group assumes in the project, and the attributes that an individual should possess to contribute effectively in that particular position.

The history of the implementation of the SSHAC guidance has led to a number of innovations and improvements in the ways that the SSHAC concepts are implemented in actual projects. Often, in fact, the implementation approaches have been customized and tailored to address the specific issues of importance to the particular study. This continual improvement and evolution of implementation approaches, as well as customizing the approaches for project-specific applications, is commendable and should be encouraged. At the same time, there is a need to describe the minimal requirements that allow one to call a particular study a Level 3 or 4 SSHAC project. If these requirements are met, embellishments and project-specific enhancements can be employed. The essential steps in Level 3 and 4 studies are listed below and described in detail in this report.

Essential Steps in SSHAC Level 3 and 4 Studies

1. Select SSHAC Level.
2. Develop project plan.
3. Select project participants.
4. Develop project database.
5. Hold workshops (minimum of three).
6. Develop preliminary model(s) and Hazard Input Document (HID).
7. Perform preliminary hazard calculations and sensitivity analyses.

8. Finalize models in light of feedback.
9. Perform final hazard calculations and sensitivity analyses.
10. Develop draft and final project report.
11. Perform participatory peer review of entire process.

The original SSHAC guidelines give only sparse guidance on practical issues of structural organization and management of a major SSHAC project, but it is of clear importance to ensuring a successful outcome. The recommendations made here are based on experiences of what has and has not worked well in practice. These recommendations may need to be adapted to the requirements and the context of each project. Roles and project organizational structures are defined and recommended for:

- Project sponsor.
- Project manager.
- PPRP (Participatory Peer Review Panel).
- Project TI or TFI Lead (Technical Integrator or Technical Facilitator/Integrator).
- Database management team.
- Specialty contractors.
- Resource experts.
- Proponent experts.
- TI leads.
- TI teams.
- Hazard calculation team.

A PSHA is always conducted for a specific purpose, ultimately linked to mitigation of earthquake risk to engineered structures or facilities. Therefore, from the very outset of the project at the planning phase, we strongly recommend engagement with the sponsor to define the required deliverables. This will usually necessitate dialogue with managers from the sponsoring organizations as well as with those who will make use of the output from the hazard calculations. A common use of SSHAC Level 3 and 4 hazard studies is to contribute to licensing or safety evaluations of safety-critical facilities. Therefore, a great advantage can be gained if the NRC (or any other relevant regulatory body or bodies) follow the entire process, primarily by attending the workshops as observers. Although observers in a SSHAC workshop are precluded from the technical discussions, we suggest that organizers allot some time at a specified time at the end of each day or each workshop to open the floor to questions and comments from observers. In such a context, the regulator could provide feedback, raise concerns, or ask for points of clarification. In two recent NRC-sponsored SSHAC Level 3 Projects (i.e., the Central and Eastern United States Seismic Source Characterization [CEUS SSC] project and the Next Generation Attenuation Relationship for Eastern North America [NGA-East] project), staff also participated in the conduct of the study.

A requirement of the PPRP is to submit written reports after each workshop and after issuance of the final draft and final PSHA reports; it is highly desirable that these be consensus documents reflecting the views of the Panel as a whole. Every effort should be made to avoid arriving at the end of the project with a "minority view" among the PPRP members. This makes the role of PPRP Chairman very important, and the individual holding this position needs to be well regarded by their colleagues and should possess the personal qualities to encourage constructive interaction, facilitate effective dialogue, and resolve differences of opinion. The PPRP Chairman should be someone with extensive experience with PSHA projects and review panels. The PPRP Chairman also should be well versed on the SSHAC process.

The imperative to capture the full range of the integrated distribution (i.e., the full range of technically defensible interpretations) should not lead evaluator experts to include alternatives in their models only to convey the impression of broad capture of epistemic uncertainty. Only truly defensible models and parameter values should ever be included in the logic-tree; assigning very small branch weights to models that the evaluator believes to be completely unsupported is not appropriate. The objective of effectively capturing the integrated distribution should remain paramount, and the evaluator should be encouraged to avoid any *post facto* alterations to his or her logic-tree strictly for the sake of simplifying the hazard calculations. This could compromise the expert's ability to speak to that model and thus their sense of ownership of the model. At the same time, at the time of assigning weights to branches, an evaluator can reap clear and obvious benefits if the structure of the logic-tree is kept simple and important elements have not been neglected.

In the absence of any realistic means of proving so, confidence that the CBR of the TDI has been captured comes from the structure and rigor of the SSHAC assessment process itself and the confirmation from the PPRP that the project conformed to the requirements of the process. The transparency and thorough documentation of the process also provides confidence. The essential steps outlined above define the basic standard that is expected to lead to assurance that the goals of the SSHAC process have been met.

A SSHAC Level 3 or Level 4 PSHA will invariably be divided into subprojects primarily to reflect the fact that the SSC and Ground Motion Characterization (GMC) elements of the hazard input require different expertise and, consequently, involve different assessments. Because all the separate components of a PSHA study are ultimately combined into a single logic-tree that defines all of the hazard calculations, it is essential that all the elements of the hazard model are compatible with one another and are defined using consistent parameter definitions.

It is important to note that, with the exception of motivational biases, the common cognitive biases (some of which are touched on above) are inherent to all expert judgments and are not deliberate. They are simply the way that we commonly process information and offer our technical judgments. Fortunately, studies have shown that the most effective way of countering cognitive biases is simply to make the experts aware that the biases exist and to encourage the experts to counter them. For example, the TI Lead or TFI can counter overconfidence by probing the limits of an expert's expression of uncertainty to ensure that the full range is being provided. The bias defined as "availability"[1] can be countered by asking the expert for the technical reasons that he or she prefers a particular model or considers other models less credible. The presence of ignored or unstated conditioning events can be brought to light by asking the expert for all assumptions that went into a particular expert assessment. One recommendation for avoiding some of the problems of anchoring expert assessments is to display only normalized hazard results during the evaluation phase of the project.

Conducting SSHAC Level 3 or 4 studies to develop community-based SSC and GMC models for an entire region or country can offer many advantages. Duplication of work on the earthquake catalogue and the ground-motion models can be avoided. Better use can be made of the available resources, particularly the limited pool of experts. If hazard assessments are required at several sites and the number of available experts is limited, then this approach also can save considerable time if the alternative is several separate site-specific studies conducted in series.

---

[1] Availability bias is an unrecognized tendency of decisionmakers to give preference to recent information, examples that can easily be brought to mind, and specific acts and behaviors that they personally observed.

Because a regional model could commonly define many elements of the input to each site-specific PSHA, each study can be expected to produce more stable and consistent results, and the bases for the individual hazard assessment are less likely to be challenged. Such community-based approaches to developing SSC and GMC models for an entire region can be implemented by the end users pooling resources to conduct these projects that would produce commonly owned results. Individual site owners can then adopt the regional SSC and GMC models and apply them to site-specific assessments, refining them for local seismic sources and including site response modifications as required.

ACKNOWLEDGMENTS

The authors wish to acknowledge and extend appreciation to some of the many individuals who contributed to the development of this document. Drs. Kevin Coppersmith and Julian Bommer were deeply involved in bringing actual project experience from the conduct of several Senior Seismic Hazard Analysis Committee (SSHAC) projects into the considerations of implementation guidance given in this document. Their advice and perspectives provided a firm basis for many of the recommendations included herein.

Dr. Norman Abrahamson provided early important input to U.S. Nuclear Regulatory Commission (NRC) staff related to the development and implementation of the SSHAC practical guidelines research program that resulted in this report. Dr. Tom Hanks, Dr. Dave Boore, and Ms. Nicole Knepprath of the U.S. Geological Survey (USGS) Menlo Park, organized the highly successful USGS/NRC joint workshops investigating 15 years of experience with Probabilistic Seismic Hazard Analysis (PSHA) using the SSHAC guidelines and produced an excellent USGS Open File report (also co-authored by Drs. Abrahamson and Coppersmith) summarizing the workshop outcomes. Dr. Cliff Munson of the NRC contributed important perspectives in the methods appropriate for update of SSHAC-informed studies. Dr. Rasool Anooshehpoor of the NRC assisted in updating many of the figures in the document. Dr. Scott Stovall and Mr. Robert Roche-Rivera provided valuable assistance in preparation of the final report.

The participants of the USGS/NRC workshops provided open, honest, and thoughtful comments and insights that were invaluable in determining what works, what does not, and how to improve the state of practice in the future. Their many significant contributions have been incorporated throughout this report and in many cases form the basis for the pragmatic advice provided. Thanks is due Jeffrey Kimball who continually asked workshop participants to think about what a true community model would look like, thereby bringing us closer to achieving that goal in the future. A special note of appreciation also is extended to the leadership and participants of the PEGASOS project in Switzerland who undertook the first international SSHAC Level 4 project and generously shared their unique perspective related to challenges and opportunities for studies outside the United States.

Special thanks and appreciation to Drs. Robert Budnitz, Jeffrey Kimball, Stephen McDuffie, Richard Quittmeyer, Leon Reiter, Frank Scherbaum, and Aybars Gurpinar who performed a review of this document in an early draft form. Their insights and input helped to form the document and have been incorporated throughout. Dr. Mike Lee of the NRC performed a review from the NRC historical perspective.

Last but not least, special thanks and appreciation to the Senior Seismic Hazard Analysis Committee itself: Drs. Robert J. Budnitz, George Apostolakis, David M. Boore, Lloyd S Cluff, Kevin J. Coppersmith, C. Allin Cornell, and Peter A. Morris. Their 4-year effort produced guidance and a conceptual framework that has had significant positive impact on the state of practice in probabilistic hazard assessments globally.

ABBREVIATIONS, ACRONYMS, AND PARAMETERS

A	Rupture Area
ACRS	Advisory Committee for Reactor Safeguards
AEC	Atomic Energy Commission
AFOE	Annual Frequency of Exceedance
AI	Arias Intensity
ANS	American Nuclear Society
ANSI	American National Standards Institute
ASCE	American Society of Civil Engineers
ASCE/SEI	American Society of Civil Engineers/Structural Engineering Institute
BTP	Branch Technical Position
DWM	Division of Waste Management
CAV	Cumulative Absolute Velocity
CBR	Center, Body, and Range
CEUS	Central and Eastern United States
CEUS SSC	Central and Eastern U.S. Seismic Source Characterization (for Nuclear Facilities Project)
CFR	Code of Federal Regulations
COL	Combined Operating License
CRWMS M&O	Civilian Reactor Waste Management System Management and Operations
DOE	Department of Energy
DSHA	Deterministic Seismic Hazard Assessment
ε	Number of standard deviations from the logarithmic mean
EE	Evaluator Experts
EPRI	Electric Power Research Institute
EPRI-SOG	Electric Power Research Institute – Seismic Owners Group
ESP	Early Site Permit
f	Frequency
FSAR	Final Safety Analysis Report
k	Flexural stiffness
κ	High-frequency motion attenuation parameter
λ_{min}	Annual rate at which the target PGA value is exceeded
G	Acceleration of gravity
GIS	Global Information System
GMPE	Ground Motion Prediction Equation (also known as an attenuation relationship)
GM	Ground Motion
GMC	Ground Motion Characterization
HID	Hazard Input Documents
HSK	Swiss Federal Nuclear Safety Inspectorate (now known as ENSI)
$H(x)$	Heavyside step function
IPEEE	Individual Plant Examination for External Events
ITC	Informed Technical Community
JMA	Japanese Meteorological Agency
LLNL	Lawrence Livermore National Laboratory
μ	Rigidity of the crust (usually taken as 3.3×10^{10} N.m^{-2})
μ	Median expected value
m	Oscillator mass
M_b	Body-wave magnitude (also noted as m_b)

MCE	Maximum Credible Earthquake
MECE	Mutually Exclusive and Collectively Exhaustive
M_L	Local magnitude
M_{max}	Largest earthquake that is considered possible within a particular seismic source (also noted as m_{max})
M_{min}	Lower limit of magnitude (generally considered to reflect the smallest earthquake considered of engineering significance; also noted as m_{min})
MMI	Modified Mercalli Intensity scale
M_o	Seismic moment
M_w	Moment magnitude
M_s	Surface-wave magnitude
n	Number of earthquakes in a Gutenberg-Richter relationship
NAGRA	Nationale Genossenschaft für die Lagerung radioaktiver Abfälle (Switzerland National Cooperative for the Disposal of Radioactive Waste)
NPH	Natural Phenomena Hazards
υ_i	Annual rate of earthquakes of mi and greater in a seismic source
$\upsilon_{m_{min}}$	Annual rate of earthquakes of magnitude m_{min} and greater
NGA	Next Generation Attenuation Relationship
NGA-East	Next Generation Attenuation Relationship for Eastern North America
NNR	National Nuclear Regulator of South Africa
NRC	U.S. Nuclear Regulatory Commission
NUREG	U.S. NRC Nuclear Regulation Report
NUREG/CR	U.S. NRC Nuclear Regulation Contractor's Report
PEGASOS	Probabilistische Erdbeben-Gefährdungs-Analyse für KKW-StandOrte in der Schweiz (Probabilistic Seismic Hazard Analysis for Swiss Nuclear Power Plant Sites)
PGA	Peak Ground Acceleration
PGV	Peak Ground Velocity
PM	Project Manager
POC	Project Oversight Committee
PPRP	Participatory Peer Review Panel
PRA	Probabilistic Risk Assessment
PSHA	Probabilistic Seismic Hazard Analysis
PTI	Project Technical Integrator
PTFI	Project Technical Facilitator Integrator
PVHA	Probabilistic Volcanic Hazard Analysis
PVHA-U	Probabilistic Volcanic Hazard Analysis Update
QA	Quality Assurance
r_{epi}	Epicentral distance
r_{hyp}	Hypocentral distance
r_{jb}	Joyner-Boore distance (also noted as r_{JB})
r_{rup}	Rupture distance
σ	Standard deviation
σ_σ	Standard deviation of the standard deviation
σ_{ss}	Single station uncertainty (sigma)
σ_μ	Standard deviation of the median
S_a	Spectral acceleration
SDC	Seismic Design Category
SDOF	Single Degree of Freedom
SNL	Sandia National Lab

SSC	Seismic Source Characterization
SSCs	Structures, Systems, and Components
SSE	Safe Shutdown Earthquake ground motion
SSHAC	Senior Seismic Hazard Analysis Committee
TDI	Technically Defensible Interpretations
TFI	Technical Facilitator Integrator
TI	Technical Integrator
TIP	(SSHAC) Trial Implementation Project
t_{max}	Duration of an earthquake recording
u	Average slip on the fault plane
UHS	Uniform Hazard Spectrum
USDOE	United States Department of Energy
USGS	United States Geological Survey
USNRC	United States Nuclear Regulatory Commission
V/H	Vertical-to-Horizontal ground motion ratio
V_s	Shear-wave velocity
$V_{s,30}$	Average shear-wave velocity over the uppermost 30 meters of a geologic column
WASH-1400	WASH-1400 is a document now called NUREG-75/014
WIPP	Waste Isolation Pilot Project
Z_{TOR}	Depth-to-top-of-rupture

1. INTRODUCTION

In 1997, the U.S. Nuclear Regulatory Commission (NRC) ssued NUREG/CR-6372 entitled, *Recommendations for Probabilistic Seismic Hazard Analysis: Guidance on Uncertainty and the Use of Experts*. The document was the culmination of 4 years of deliberations by the Senior Seismic Hazard Analysis Committee (SSHAC) regarding the manner in which the uncertainties in probabilistic seismic hazard analysis (PSHA) should be addressed using expert judgment. The document describes a formal process for structuring and conducting expert assessments that has come to be known as a "SSHAC process," and the recommendations made in the report are referred to as the "SSHAC guidelines." To account for different project needs and projects undertaken in different regulatory contexts, the SSHAC report describes four 'Study Levels" that define the processes and complexity of the recommended project activities. Study Levels 3 and 4, referred to hereafter simply as SSHAC Levels 3 and 4, are the most complex and involve the greatest amount of effort.

This introduction describes the purpose of this NUREG report and its context in the framework of existing guidance and experience gained from the conduct of probabilistic hazard studies over the past 15 years. Also described are the relationship of this document to the original SSHAC report, the scope and limitations of this guidance, and an overview and roadmap of this report.

1.1 Purpose of this NUREG

This NUREG serves two primary purposes—it provides (1) additional levels of detail on topics related to the implementation of SSHAC processes beyond those provided in the original SSHAC report (NUREG/CR-6372), particularly for Level 3 studies, and (2) additional guidance in the implementation of Level 3 and 4 studies in light of experience gained from past SSHAC projects. Over the past 15 years, several SSHAC Level 3 and 4 studies have been conducted, thus leading to an expanded "database" of experience in the intricacies of carrying out actual SSHAC projects.

The original SSHAC study was motivated by the need for a consistent methodology that could be used for a wide variety of applications—not confined exclusively to nuclear facilities—and that would consequently lead to increased consistency in the hazard results obtained from different studies. As described in the Executive Summary to the SSHAC report (provided in Appendix A), the technical environment in which a PSHA is undertaken is one involving considerable uncertainty. The same claim can be made for other hazard analyses that involve sparse data and the potential occurrence of rare, high-consequence natural events such as volcanism and tsunami hazards. Given the regulatory context of the committee's work, and the virtual absence of preexisting guidance related to processes for undertaking PSHA studies, the SSHAC guidance addresses the process issues at a high level and supports each process step with a discussion of its methodological underpinnings. For example, the nine "key procedural points" given in the Executive Summary of the SSHAC report focus on the issues of defining expert roles, the unique role of the Technical Integrator and the Technical Integrator/Facilitator (defined and discussed in Section 3.6.4 and 3.6.6), the evaluation and integration process and the need for process and technical participatory peer review. Rather than providing detailed implementation steps, the SSHAC report outlines the important components of a SSHAC process and makes suggestions for process-related steps that were believed to be useful in achieving these components.

Although the SSHAC report does acknowledge the various constituencies who might be conducting a PSHA, defines four "Study Levels"[2], and identifies the decision factors that should be considered in selecting a SSHAC Level, it focuses heavily on the processes for a Level 4 study. This was in large part because the higher study levels (i.e., levels 3 and 4) are designed to provide the higher levels of "regulatory assurance" (defined in Section 3.3) needed for nuclear facilities, which was the principal interest of the SSHAC sponsors. Also, the focus on Level 4, which involves formal assessments by panels of experts, was partially motivated by the desire to frame the SSHAC process in the context of previous expert-based studies. Many of those studies also had their origins in applications for nuclear facilities, particularly for risk analyses. The SSHAC report does not address the approaches and processes for conducting Level 1-3 studies in much detail other than to note that they are led by Technical Integrators (without the need for the Facilitator role required for a panel of experts) and that the essential SSHAC goals remain the same. In this sense, the SSHAC report fails to distinguish between Levels 1, 2, and 3 in terms of important activities and processes.

Over the past 15 years, each SSHAC project has provided an opportunity to apply the SSHAC concepts and to evolve or enhance the implementation approaches. For example, the various expert roles in a SSHAC process are important (e.g., evaluator, proponent, resource experts) and practical approaches to identifying and reinforcing those roles throughout the course of a project have evolved. Likewise, structured expert interactions essential to Level 3 or 4 studies and effective approaches for determining timing, duration, and content of workshops can now be defined more completely, based on project experience.

Also, recent years have seen increasing interest in the SSHAC process as a method of achieving high levels of regulatory assurance for the safety evaluation and licensing of nuclear facilities. Owners of existing nuclear facilities are seeking highly credible methods for demonstrating that members of the technical community have been consulted in the hazard studies used to evaluate the safety of their facilities. License applicants are also seeking structured, expeditious, and approved approaches to characterizing hazards that will be viewed as acceptable to the NRC (or other regulators). Regulators are looking for high levels of assurance that hazard studies have properly captured the knowledge and uncertainties of the technical community and that the technical assessments are transparent and fully documented

The audience for this document is varied and includes sponsors, regulators, hazard analysts, technical experts, and reviewers. Hazard studies to which the guidelines can be applied include probabilistic assessments of ground motions, fault displacement, volcanism, and tsunami hazards. Sponsors of probabilistic hazard analyses are ultimately responsible for commissioning these studies and therefore need to understand the essential elements of a SSHAC study. In addition, a clear understanding of the essential elements of SSHAC studies conducted at particular SSHAC Levels to assess goals is of clear benefit to regulators—such as NRC staff—who will be responsible for reviewing a hazard study being prepared to support a license application or a safety review for a nuclear facility. Hazard analysts, technical experts, and reviewers can use this document to understand the details of the implementation processes and the roles that various project participants are expected to play. All of the participants in a licensing process, including the users of the hazard information, may benefit from reading this document to understand the uncertainties associated with the hazard analysis and the manner in which the views of the larger technical community have been considered.

Valuable input to this NUREG comes from a series of workshops sponsored by the NRC in 2008 (Hanks et al., 2009) that were attended by practitioners of PSHA and probabilistic volcanic

---

[2] Study levels are termed SSHAC Levels or SSHAC Study Levels in this document

hazard analysis (PVHA) from both the private and public sectors. These workshops (1) identified and discussed the SSHAC Level 3 and 4 projects conducted over the past 2 decades, (2) described the "lessons learned" from these case studies in terms of what works well and not so well among various project implementation approaches, (3) drew conclusions regarding the various process elements that have proven to be beneficial in actual practice, and (4) investigated how and when the results of studies should be updated or replaced. In addition, the workshops identified and discussed problems and difficulties arising from past approaches and methods. With the information gained at these workshops as a backdrop, this NUREG further defines the SSHAC goals and methodology and provides greater specificity regarding the recommended approaches for conducting Level 3 and 4 studies.

1.2 Relationship of this NUREG with the SSHAC Guidelines

The NRC intends this document to complement rather than replace the original SSHAC report (NUREG/CR-6372). The SSHAC document presents recommendations regarding methodology that were made largely in the abstract without the benefit of significant experience in many areas that would allow for specific guidance. As a result, the recommendations provide a very useful framework but are generally at very high level. This document is intended to fill in that framework with details (necessarily) missing in the original SSHAC guidelines. In addition, the SSHAC guidelines devoted most of the effort to discussing SSHAC Level 4 studies and provided very little guidance regarding the specific approaches that would be appropriate for Level 3 studies. For this reason, this document—and the recent work supporting it—has focused a significant amount of effort on developing and describing the appropriate methods and approaches for a Level 3 study.

In the context of actual practical project applications, the conceptual underpinnings of the original SSHAC guidelines remain strong. In this NUREG, we offer some clarification and augmentation in terms of terminology with respect to the original guidelines. Some terms employed in the guidelines were appropriate given the contemporary usage when the guidelines were written. They have since been defined in new ways. For example, we recommend against use of the term "expert elicitation" when describing a SSHAC process—even for a Level 4 study that involves the use of multiple experts. This is because the decision analysis community has continued to use the term to describe processes that are quite different from those that define a SSHAC process. As another example, details of the "two-step" *evaluation* and *integration* process described in the SSHAC guidelines are now clarified both in terms of the means by which each of these steps are actually undertaken and in terms of the evidence provided in project documentation to demonstrate that the steps have been faithfully followed.

1.3 Scope and Limitations of These Guidelines

The original SSHAC study resulted in guidelines that describe and provide for the basic philosophy of a properly conducted hazard project. This NUREG follows the SSHAC guidelines with the same level of "prescription" and offers additional levels of specific guidance based on experience gained from actual applications and projects. Strong recommendations in this document will be distinguished by the use of the term "should" and lesser recommendations by the use of the term "could." It is recognized that innovative approaches to achieving the SSHAC goals will continue to be developed in the future and that project-specific refinements to the approaches discussed here may be appropriate. However, the application of the guidance given in this document will most likely lead to greater stability and longevity of the hazard assessment being made. Likewise, higher levels of regulatory assurance are likely to be gained with careful and conscientious application of this guidance. Proper application of the guidance given in this document is not guaranteed to lead to acceptance of the hazard results by the NRC or any other regulator. However, use of these guidelines should lead to reduced review times by regulators and higher likelihoods of acceptance of the hazard results.

1.4 Overview and Roadmap of this Document

This report follows a format designed to lead the reader from introductory materials that provide a framework for understanding the historical context of SSHAC, to detailed implementation guidance, to discussions and recommendations for application of the guidelines within the context of future studies. The NRC developed the report by first carefully reviewing the issues of importance to hazard analysis and existing SSHAC guidance. Then the NRC reviewed the projects conducted during the 13 years since the document was first issued and developed guidance and recommendations that take advantage of the practical knowledge gained from actual applications.

Following this introduction, which establishes the purpose of the document and its relationship to the existing SSHAC guidance, Chapter 2 is designed to provide the history of the use of multiple experts to address uncertainties for hazard and risk analyses. The SSHAC process finds its roots in earlier structured approaches that were used for capturing expert judgments. The goal of this chapter is to allow the reader to understand the context of studies that motivated the original SSHAC guidance and to describe the evolution of the processes used in studies that have occurred since the SSHAC guidelines were issued. This history provides the fundamental basis for the implementation guidance given in this NUREG.

Chapter 3 presents key SSHAC concepts that are the underpinnings of all SSHAC projects and defines them to help the reader understand the firm conceptual basis for the methods recommended in this document. This chapter also provides descriptions of the key roles and responsibilities of the various participants in a SSHAC process. These roles are an important part of what sets SSHAC processes apart from other studies.

Chapter 4 provides a succinct description of the essential steps that comprise SSHAC Level 3 and 4 studies. The goal of this chapter is to define the minimum required activities in each step that must be conducted if the claim is to be made that a hazard study is a SSHAC Level 3 or 4 study. In addition to the required steps, the section also provides suggestions for ways that the various process steps can be carried out and documented.

Chapter 5 draws on the experience of past studies to give practical guidance on various approaches to organize and implement a SSHAC project. It addresses a number of aspects including suggested project organizational structure, roles of the sponsor and the peer review panel, and community-based studies versus individual studies. Chapter 5 also discusses the interrelated issues of cognitive bias, demonstrated capture of the full range of epistemic uncertainty, and dealing with large and complex logic-trees. This chapter's recommendations are designed to allow the reader to take advantage of the lessons learned from past studies.

Acknowledging that a hazard study represents a snapshot in time and that the technical community will develop new data, models, and methods after the study is conducted, Chapter 6 discusses replacing, revising, and refining a hazard study. This chapter gives a discussion of the ways that a SSHAC process addresses NRC regulations and regulatory guidance. It also considers alternative approaches for assessing whether or not a hazard study needs to be replaced, whether new data and findings are significant and would require revision of the study, and whether site-specific refinements to a regional hazard model are desirable.

Chapter 7 provides a summary of the key findings and recommendations.

Appendix A provides the Executive Summary for the original SSHAC report as a means of summarizing the key findings and recommendations of the original SSHAC study.

Appendix B provides an overview of probabilistic hazard analyses. Its goal is to provide a basic explanation of the issues that are important to hazard analyses, define the uncertainties that most affect hazard results, and present the types of hazard studies that can benefit from a SSHAC approach.

Appendix C provides a perspective on this NUREG from the Chairman of the original Senior Seismic Hazard Analysis Committee, Dr. Robert Budnitz.

2. HISTORY OF MULTIPLE-EXPERT HAZARD ASSESSMENTS

The purpose of this section is to provide an historical context for the guidance being provided in this document. The Senior Seismic Hazard Analysis Committee (SSHAC) report was written in response to an evolution of expert assessment methodologies that had been used to conduct probabilistic risk analyses during the previous 3 decades. The methodological guidance provided in the SSHAC report was intended to build on the lessons-learned from these previous studies and, specifically, to arrive at processes that would avoid the problems that had plagued previous studies.

In the same way, this NUREG seeks to enhance the existing SSHAC guidance by addressing the strengths and inadequacies identified from the experience gained since the time that the SSHAC report was issued in 1997. In this sense, the recommendations also are part of the evolution of guidance in this area by the U.S. Nuclear Regulatory Commission (NRC) and other groups. Accordingly, this section of the report summarizes the significant studies and case histories that led to the development of the SSHAC guidance and then further recaps the SSHAC-based assessments that have occurred up to the present time.

As will be illustrated in the subsequent discussions, the NRC has a long history of addressing uncertainties in seismic hazard and risk studies. Moreover, because many of the technical issues that drive these assessments are not readily characterized using empirical data, The NRC has explored the formal use of expert judgment to supplement empirical knowledge. Because of the need to use expert judgment, the probabilistic hazard community explored the decision analysis field of "expert elicitation" early on. Indeed, the terminology of "expert elicitation" has permeated the probabilistic seismic hazard literature for many years and, in fact, the SSHAC report calls the process a "formal, structured expert elicitation process." However, as discussed in Section 3.5, the development of the SSHAC guidance represented a departure from classic expert elicitation processes. Now, with considerable experience from multiple projects, the SSHAC process has reached a point in its evolution that requires a distinction be made concerning classic expert elicitation. This is because the SSHAC methodology includes attributes that are not consistent with the current definition of an expert elicitation process. Such attributes include a focus on identifying and evaluating the views of the larger technical community, dissemination and sharing of common databases, interaction of experts in workshops, feedback and challenge of experts by their colleagues, etc. With this realization in mind, this document makes a distinction between SSHAC processes and classic expert elicitation processes. Nevertheless, as will be discussed below, the history of expert assessments leading to SSHAC has its roots in classic expert elicitation.

To assist the summary of the historical context for this NUREG, Table 2-2 (found at the end of the chapter) presents several notable multi-expert probabilistic hazard studies and guidance studies conducted over the past 3 decades. Table 2-2 is not an exhaustive listing of all such studies but is representative and intended to show the evolution of studies that led to the development of the SSHAC report in 1997 and the studies that have followed. The post-1997 studies (and those that occurred during the 4 years that SSHAC was being developed) include only those studies that followed SSHAC Level 3 or 4 processes. No doubt, during the same time, many more studies were conducted using SSHAC Level 1 and 2 or non-SSHAC approaches, but those are not the focus of this NUREG.

2.1 Multiple-Expert Assessments Before SSHAC

Early procedural guidance for nuclear power plants, such as Appendix A to 10 CFR Part 100, called for the use of deterministic approaches to define the design basis ground motions at nuclear power plant sites. These studies did not explicitly take into consideration the likelihood of occurrence of the design basis motions nor the recurrence rate of the earthquakes that predominantly contributed to the predicted ground motions. As a result, probabilistic seismic hazard studies conducted subsequently showed that the median annual frequency of exceeding the design basis ground motions (safe shutdown earthquake ground motion [SSE]) for existing nuclear power plants spans a range of nearly two orders of magnitude[3] (USNRC, 1997). The early deterministic studies also did not have a formal mechanism for acknowledging and incorporating uncertainties in key seismic source characteristics into the analyses. Instead, notions of "conservatism" were used to identify maximum credible earthquakes to define the SSE ground motions. Of course, in the absence of quantified uncertainties, the degree of conservatism could not be quantified, and discussions between license applicants and the NRC often centered around appropriate levels of conservatism in the SSE inputs. In the absence of quantified risk information, very little discussion could occur regarding the hazard and risk significance of the inputs and consistency was difficult to achieve.

Well before frameworks had been developed for formally eliciting and incorporating expert judgments, regulators looked to consideration of all sides of technical issues as a means of informing decisions and achieving regulatory stability. For example, the NRC has conducted research and commissioned independent parallel studies as a means of gaining insight into the knowledge and uncertainties associated with key technical issues. Likewise, the licensing process in the United States provides opportunities for intervention by groups representing the public, and the public hearing process is designed to allow experts with differing views to express themselves before a licensing board. In making its decisions, the NRC has concluded that a consideration of the full range of views within the expert community provides a deeper understanding of the uncertainties, particularly for decisions regarding rare, high-consequence events.

Studies using formally-elicited expert opinion started at the RAND corporation as early as the 1950s and 1960s (Dalkey and Helmer, 1963; Dalkey, 1967 and 1969) and were rapidly enhanced by consideration of probabilistic approaches to analyzing the gathered responses as developed by researchers at the Stanford Research Institute (Spetzler and Staël von Holstein, 1975) and Schlaifer, 1959. Earth science-based applications of formal expert elicitation, particularly for addressing issues related to the management of high-level radioactive waste, were found to be useful in evaluating areas of epistemic uncertainty (EPA, 1985; NRC, 1991b). By the early to mid-1970s, the concept of expert elicitation was being investigated specifically for use in seismic hazard assessment applications as a way of addressing the statistical uncertainties in the geologic data on which the hazard estimates themeselves were based (Okrent, 1975).

The early studies of reactor safety (e.g., the WASH-1400 study [USNRC, 1975] and the study described in NUREG-1150 [USNRC, 1991a]), began the explicit use and incorporation of expert judgment as a means of extending existing knowledge. The existing knowledge base ranges from observed data based on actual experience to models of rare and highly uncertain events for which data are not available. For example, experts in these early safety studies were provided with certain assumptions or scenarios—such as a "loss of offsite power" or a "loss of

---

[3] See for example, Figure B-2 of Appendix B in USNRC 1997.

cooling accident"—and were asked to provide their assessments of the accident sequences that would result and the probabilities associated with those accident sequences. Appropriately, studies of risk were probabilistic to account for two important aspects: (1) the probability or frequency of occurrence of various elements of the analysis, and (2) the uncertainties in the models and parameter values that are used to define the risk. The early studies focused primarily on developing the technical framework for assessing the risk such as the basic elements of the model (e.g., event trees) and the probabilistic assessments (e.g., probability of failure of structures, systems, and components [SSCs]). Less emphasis was placed on the manner in which expert judgments were elicited or on formalizing the process used for quantification of uncertainties. Likewise, the early studies devoted very little effort to quantifying the likelihood of the *hazard* (ground motion input) but focused more on the consequences of the engineered system, given a level of ground motion.

Moving through the late 1970s into the early 1980s, the technical community involved in seismic hazard analyses developed an increasing appreciation of the importance of uncertainties in the analyses. This is especially true for assessments of hazard for rare events, where the models and probability distributions based on empirical observations must be extrapolated in space and time to annual frequencies that are very low relative to the observed data. At the same time, members of the hazard and risk community began to adopt the structure and formalism of "expert elicitation" processes from the decision analysis community. Along with the adoption of those techniques, concern began to arise regarding the possible impact of "expert elicitation issues" such as those given in Table 2-1.

Table 2-1. Example Expert Elicitation Issues Identified During the 1970s and Early 1980s

Topic	Example Expert Elicitation Issues
Breadth and balance of an expert group	• The degree to which any group of experts provides a balanced and complete sampling of the larger community
Independence	• The ability of an expert to provide his/her own views and not those of their peers or agency
Elicitation protocols	• Whether interactions among experts should be encouraged or prohibited • Assessments made individually or in a group setting
Cognitive biases	• Training to avoid anchoring, underestimation of uncertainty, etc.
Consensus	• Required, encouraged, or not required across a panel of experts
Aggregation	• Combination rules for diverse expert assessments • Mechanical, behavioral • Equal versus unequal weights
Documentation	• Capturing expert reasoning and thought processes, or just the outputs

Despite the fact that seismic design criteria for nuclear power plants had been defined using deterministic approaches, the NRC commissioned studies in the early 1980s called the Systematic Evaluation Program (SEP) to examine the probabilistic risk at the oldest nuclear power plants. As part of the SEP, consideration was given to the use of expert opinions to characterize the uncertainties in seismic hazard and fragilities. For example, NUREG/CR-1582 (Bernreuter and Minichino, 1982) developed estimates of the seismic hazard at the sites of the nine oldest nuclear power plants east of the Rocky Mountains. The study included a consideration of "the formal elicitation of expert opinion to obtain a subjective representation of parameters that affect seismic hazard at the nine SEP facilities." In NUREG-0967 (Reiter and Jackson, 1983), the NRC Geosciences Branch evaluated the probabilistic estimates presented in NUREG/CR-1582 and compared and modified them to take into account deterministic estimates. NUREG-0967 presented the NRC Geosciences Branch's first approach to utilizing complex state-of-the-art probabilistic studies in an area where probabilistic criteria had not yet been set and where decisions for specific plants had been previously made in a nonprobabilistic way. The SEP also resulted in probabilistic seismic hazard results at the U.S. Department of Energy (DOE) sites (Coats and Murray, 1984).

Perhaps the most dramatic and important revelations regarding the significance of expert assessment methodologies emerged when parallel regional Probabilistic Seismic Hazard Analyses (PSHAs) were conducted for central and eastern U.S. nuclear power plant sites. The Electric Power Research Institute-Seismicity Owners Group (EPRI-SOG, 1988, 1989) and Lawrence Livermore National Laboratory (LLNL) (Bernreuter et al., 1989) studies were both conducted using multiple experts, and both studies were conducted mindful of the importance of uncertainties. However, the processes used to conduct the studies were quite different.

The LLNL study consisted of two panels of experts (one for SSC and one for ground motions). The experts were each asked to identify any data that they felt would be important for their assessments, and they were provided with those data. Written questionnaires were provided to the experts that enumerated the desired inputs, and support from the team leaders was provided (in writing, phone calls, and meetings with each expert) to assist the experts in their assessments. The experts did not meet in workshops and were not encouraged to discuss their assessments with other panel members. To maintain anonymity, expert assessments were not attributed to the experts by name. The team leaders reviewed the experts' initial assessments and asked questions of clarification to each expert to ensure accuracy. In addition, limited feedback was provided regarding the implications of the assessments to the hazard results.

In the EPRI-SOG study, a project-wide database was developed that was made available to all experts. The study focused principally on seismic source characterization issues (ground motions were handled by the project team), and the experts were assembled into teams that each included a seismologist, geologist, and geophysicist. Workshops were held to identify data and alternative hypotheses. To review the expert assessments; elicitation sessions were held between the project team and each expert team; and limited feedback regarding the implications of the expert assessments was provided.

Although the methodological differences used in the two studies were known at the time the studies were conducted, no procedural guidance existed and there was little indication that the differences would have a significant effect on the results. However, a comparison of the calculated hazard results at the 56 common sites in the central and eastern United States (CEUS) showed significant differences between the two studies (summarized in USNRC 2010, NUREG-0933, Generic Issue 194). The DOE and NRC looked into the issue and found that an important contributor to the difference was the seismicity experts' input related to lack of correlation between the recurrence parameter "a" and "b" values. This issue was the driving force behind NRC formal updating of the LLNL results as documented in NUREG-1488 (Sobel, 1994). NUREG-1488 both compares the studies and provides an updated model. Another key concern on the part of the LLNL study was the fact that a single ground motion expert provided assessments that were well outside of the range of assessments provided by the other four ground motion experts on the panel. The "outlier" expert's assessments had a significant impact on the mean estimate that was calculated based on input provided across the panel. The concern regarding the unbalanced impact of outliers on assessments was largely addressed in a followup study by LLNL that took a different approach to uncertainty characterization by developing a composite ground motion model (Savy et al., 1993). However, the differences between the LLNL and EPRI-SOG hazard estimates remained—particularly in the range of annual frequencies of 10^{-4} to 10^{-6}, which is the range of seismic hazard that typically has the most contribution to seismic risk for nuclear power plants (USNRC 2010, NUREG-0933, Generic Issue 194). Within this range, the LLNL mean hazard results were systematically higher than the EPRI-SOG results.

The NRC addressed the issue of the discrepancies between the EPRI-SOG and LLNL studies in two ways: (1) seismic risk studies were conducted as part of the Individual Plant Examination for External Events (IPEEE) to ensure that the existing plants would have adequate seismic margins against the revised hazard estimates and (2) the Senior Seismic Hazard Analysis Committee (SSHAC) was established with the purpose of examining the differences between the LLNL and EPRI-SOG studies and "resolving the differences." The IPEEE program began in 1991 and, although the SSHAC was convened beginning in 1993, it would be 4 years before the Committee's findings would be released.

In the early 1990s, a number of multi-expert studies were conducted to evaluate hazards at nuclear facilities for licensing or to demonstrate expert elicitation methodologies. The Satsop PSHA was conducted for a proposed nuclear power plant site in western Washington State and focused on the uncertainties in the seismogenic potential of the Cascadia subduction zone (Coppersmith and Youngs, 1990). A panel of 14 experts was asked to address a series of questions and technical issues related to the geometries, maximum earthquake size, and recurrence rate of various subduction zone sources. The goal was to capture the state of knowledge and uncertainty in the technical community at that time by eliciting the judgments of a relatively large number of experts. At about the same time, Sandia National Laboratories conducted another expert elicitation project for the Waste Isolation Pilot Project (WIPP) in Carlsbad, New Mexico. The purpose of the study was to supplement the performance assessment for the repository at WIPP (Trauth et al., 1991). In that study, a relatively small panel of experts was given access to the geologic and hydrologic data developed for the site and was asked a series of focused questions regarding the performance of the repository system. Experts were asked to develop their own models of the system to arrive at a common measure of performance in terms of radionuclide concentrations at various distances from the repository at various points in time.

Two important "demonstration" projects also were conducted in the early 1990s whose purpose was to illustrate the manner in which expert elicitation methodologies could be used to address highly uncertain hazard issues. The Center for Nuclear Waste Regulatory Analyses (CNWRA) *Expert Elicitation of Future Climate in the Yucca Mountain Vicinity* study, funded by the NRC, was intended to demonstrate the manner in which a panel of experts could be used to make assessments about the future climate over the next 10,000 years in the Yucca Mountain region of Nevada (DeWispelare et al., 1993). The project was not conducted by the future license applicant at Yucca Mountain (DOE), but by a contractor to the regulator (the NRC). Thus, the focus of the study was on the *process* that could be followed, rather than on the results. In a similar vein, the Electric Power Research Institute (EPRI) sponsored a demonstration project to illustrate expert elicitation methodologies for assessing the fault displacement hazard at Yucca

Mountain in its *Earthquakes and Tectonics Expert Judgment Elicitation Project* (Coppersmith et al., 1993). Drawing on the experience gained from the EPRI-SOG project, the study placed an emphasis on holding workshops in which experts having different and alternative viewpoints on key technical issues were brought together. Both the CNWRA and the EPRI studies dealt with the expert selection process, the manner in which the judgments of experts should be obtained and assessed (e.g., in group or individual sessions), and the documentation developed by the experts to describe their assessments and their technical bases.

The mid-1990s witnessed two important efforts related to the proposed repository at Yucca Mountain—the *Probabilistic Volcanic Hazard Analysis* (PVHA) (CRWMS M&O, 1996) and the development of the expert elicitation guidance in NUREG-1563 *Branch Technical Position on the Use of Expert Elicitation in the High-Level Radioactive Waste Program* (Kotra et al., 1996). The PVHA was designed to address the potential hazard represented by the disruption of the repository by volcanism. A panel of ten experts was involved in the analysis of site-specific and analogue datasets, development of models, and characterizing uncertainties in the spatial and temporal aspects of future volcanism for the next 10,000 years. Large uncertainties accompanied many of the technical issues and technical debates among experts were encouraged in a series of workshops and field trips. Large numbers of observers, including the NRC and representatives of several Federal, State, and local oversight groups, attended the workshops.

During the time that the PVHA was being conducted, the NRC was developing NUREG-1563 (Kotra et al., 1996), which was intended to provide guidance to the DOE on the staff's expectations for the manner in which such a study should be conducted for the high-level waste program. The guidance presented a number of innovative findings, such as the distinction between an expert elicitation process, a peer review process, and less formal processes that rely on expert judgment. In the document, the term "expert elicitation" is reserved for the formal, structured process that is defined by a series of specific steps. These "components in an acceptable expert elicitation process" are consistent with the processes advocated at the time by the decision analysis community (e.g., Hora, 1993; Kaplan, 1992; Meyer and Booker, 1990; Morgan and Henrion, 1990) and include:

- Definition of objectives.
- Selection of experts.
- Refinement of issues.
- Assembly and dissemination of basic information.
- Pre-elicitation training.
- Elicitation of judgments.
- Post-elicitation feedback.
- Treatment of disparate views and aggregation of judgments.
- Documentation.

These steps, with some modification and refinement, are consistent with the basic steps advocated in the SSHAC guidance. NUREG-1563 (Kotra et al., 1996) also traced the use of expert judgment back to the earlier risk and safety studies conducted for nuclear plants and offered a vision of future guidance, such as that offered in this document:

> *"Although there are several examples of the use of expert elicitation in a nuclear regulatory context, no formal Agency guidance on this subject exists. Thus, in developing this branch technical position (BTP), the Division of Waste*

Management (DWM) staff has also drawn from previous staff experience of other NRC program offices, in the use of expert elicitation. In this regard, DWM staff has relied on certain Agency resource documents, such as: 'Risk Assessment: A Survey of Characteristics, Applications, and Methods Used by Federal Agencies for Engineered Systems'; 'A Review of NRC Staff Uses of Probabilistic Risk Assessment'; and 'Recommendations for Probabilistic Seismic Hazard Analysis Guidance on Uncertainty and Use of Experts', to help formulate its position statements. Consequently, the reader will find that this BTP is largely consistent with these other resource documents, in substance.

Subsequent to the finalization of this BTP, the staff may elect to develop guidance on the use of expert judgment in other areas of nuclear regulatory regulation.'
(p.vii)

It is important to note that during the development of NUREG-1563 (Kotra et al., 1996) the authors were observers of the Yucca Mountain probabilistic volcanic hazard assessment PVHA and in communication with those developing the SSHAC guidance (referred to by title in the quotation above). As a result, all parties were able to draw on their experience in the application of various process steps.

With an eye toward the new guidance issued by the NRC in NUREG-1563 (Kotra et al. 1996), the DOE commissioned a series of expert elicitations to address several technical issues of importance to the Yucca Mountain repository. As part of DOE's Viability Assessment, a large probabilistic risk analysis termed a "total system performance assessment" was conducted to evaluate the long-term performance of the repository system following its closure and 10,000 years into the future. Although many of the models and technical issues addressed in the performance assessment were informed by empirical data and conventional models, other inputs involved large uncertainties due to their rarity or to their extrapolation over long time periods. The issues addressed in these expert assessments were the following:

- Unsaturated zone flow: The amount and spatial distribution of groundwater flux that will percolate through the 300m of rock above the repository horizon (CRWMS M&O, 1997)

- Near-field environment and altered zone coupled effects: The thermo-hydro-chemical response of the rock and groundwater system due to the time-dependent heat output of the repository (CRWMS M&O, 1998a)

- Waste package degradation: Corrosion mechanisms and rates of penetration of the waste package for time-dependent changes in the thermo-chemical environment (CRWMS M&O, 1998e)

- Waste form degradation and radionuclide mobilization: Following waste package breach, mechanisms for waste form dissolution and geochemical mobilization, including both diffusive and advective flow (CRWMS M&O, 1998d)

- Saturated zone flow and transport: At the water table, spatial and temporal distribution of saturated zone flux down gradient to receptor at 18 km distance, including transport mechanisms such as colloids (CRWMS M&O, 1998c)

Each topic was addressed using a panel of experts, multiple workshops, interactions, interviews, and documentation very similar to that employed on the probabilistic volcanic hazard assessment (PVHA) study that had been recently completed. Significantly, unlike the PVHA study, which entailed about 2.5 years, these studies each lasted about 1 year (they were overlapped to ensure completion in 3 years). To accomplish this short time schedule, the expert panels included only four or five members, experts were required to commit to a contracted schedule, workshops were scheduled close together, and the number of technical issues addressed was limited to those required for the performance assessment.

2.2 Development of the SSHAC Guidelines

In 1993, SSHAC began its deliberations with a consideration of the procedural aspects of the EPRI-SOG and LLNL PSHAs—as well as other studies that had been conducted—and the associated pros and cons identified. Section 3.3.2.2 of the SSHAC report captures the lessons learned from these deliberations. For example, previous studies had not clearly identified the roles and responsibilities of the various participants in the study and, as a result, it was not clear who would claim "ownership" of the study after its completion. It also was observed that the lack of face-to-face interaction among experts could lead to unintended differences of opinion due different assumptions or overly narrow perspectives. Likewise, it was found that some studies lacked sufficient feedback to provide the experts with a clear idea of the implications of their models, prior to their finalization. Importantly—from the standpoint of the expert assessment process—it also was observed that the previous studies had not successfully dealt with the "outlier expert" issue. As stated in the SSHAC report:

> "Previous PSHA and multiple expert studies have dealt awkwardly or not at all with the contentious issue of "outlier experts," experts who make interpretations that are significantly different than those of the rest of the panel and that are not well supported by logic or data. Treatment of outlier experts can have a major impact on the final hazard distribution; indeed, this issue was a primary motivation for the TFI process." (NUREG/CR-6372, p. 34)

The ground motion component of the LLNL study highlighted the issues associated with the composition of expert panels and the dynamics of expert interaction. A single "outlier" assessment had a large and significant impact on the calculated results. This raised several issues to be confronted by the SSHAC in the course of its deliberations: Can goals of a study be met if expert panels consist of a collection of proponents, each with their own personal viewpoints? How many proponent experts would need to be on a panel to ensure that the views of the entire technical community are represented? How should the proponent views be aggregated such that, collectively, they represent the views of the larger community? Would interaction of the experts, and the associated challenge and defense of technical positions, provide a means of avoiding extreme outlier positions? Can or should unequal weights be applied to expert assessments during aggregation to arrive at a resulting aggregate distribution that is representative of the larger community?

Once the SSHAC delved deeply into the LLNL and EPRI-SOG studies and identified the important lessons, a decision was made to move beyond finding a resolution between the two studies. As stated in the SSHAC report:

> "Originally, some of the sponsors and participants proposed that one key objective should be to 'resolve' the differences between the LLNL and EPRI studies. However, the Committee quickly realized that the new project would be most useful if it were forward-looking rather than backward-looking - specifically, if it could pull together what is known about PSHA in order to recommend an improved methodology, rather than specifically attempting to figure out which of the two

studies was 'correct,' or which specific problems with either study were most important in affecting the study's specific results.

Therefore, although the Committee has carefully studied both the LLNL and EPRI studies (along with other past PSHAs) to obtain methodological insights, both positive and negative, we did not undertake a forensic-type examination to identify past "errors" or their implications. More broadly speaking, the Committee has attempted to draw upon the entire body of PSHA literature and experience, which is of course much more extensive than the LLNL and EPRI projects, as important as they have been." (NUREG/CR-6372, p.3)

Given the importance of the issues that emerged during the review of previous studies, the charge to the SSHAC of recommending an "improved methodology" was made in the midst of several considerations, including the following:

- The methodology would need to find a scheme that does not allow an outlier expert to unreasonably drive the mean of the aggregate distribution across all experts. This does not mean that outlier interpretations should be excluded because one does not like the results, it means that methodology must be capable of identifying and dealing with outlier interpretations.

- At the same time, the scheme would need to not suffer from the classic "sampling" problems of using a subset of experts from the larger community of experts such as: nonrandom samples of experts, experts with different levels of familiarity with the available data, experts with different motivations, dependent experts, and ack of clarity in the ownership of the aggregate distribution across all experts.

- The methodology would need to be cognizant of the problems associated with aggregation of expert assessments and to allow for both mechanical and behavioral aggregation approaches. Mechanical approaches follow a given formula or routine, and behavioral approaches rely on interaction and communication among the experts

- Any recommendations for expert assessment methodology would need to be mindful that probabilistic hazard analyses are typically conducted in the framework of contentious regulatory processes where differing views are highlighted.

- Regulatory decisions need to be made in the face of uncertainty because resources are limited. A dynamic will exist between the desire to involve large numbers of experts from the technical community in an extended debate and limited resource requirements constraining the numbers and the timeframe. Thus, the time/cost resources of the expert assessment would need to scale with the importance or risk significance.

- The recommended approaches would need to be mindful of the increased recognition of the importance of uncertainties in probabilistic hazard studies, particularly as they affect the mean estimates. Just as uncertainties are important to hazard, probabilistic external-event hazard estimates were being shown to be important to risk analyses, which was becoming an important tool for nuclear plant safety assessments.

The SSHAC dealt with a large number of the expert assessment issues by establishing and defining two important activities: evaluation and integration. Evaluation entails a number of activities including identifying important technical issues and the applicable data to address those issues, evaluating the data in terms of their quality and relevance to the assessments to be made, and facilitating interaction among the experts and members of the larger technical community to exchange viewpoints and to challenge proponents. Integration is model-building to arrive at a defensible expression of knowledge and uncertainty in inputs to the hazard model, giving due consideration to the available data and the views of members of the technical community who are not necessarily directly participating in the project. This includes the full expression of the model elements (logic-tree branches), their relative weights, and the range of credible uncertainties. As defined and explained in NUREG/CR-6372, the integration process was a new concept and provided a framework for subsequent multi-expert hazard studies.

2.3 History of Implementation of the SSHAC Guidelines

This section discusses those studies that were conducted after the issuance of the SSHAC guidelines and that had the specific goal of implementing those guidelines for SSHAC Level 3 and 4 studies.

Although completed after the SSHAC report was issued, the Yucca Mountain Project (YMP) PSHA (CRWMS M&O 1998b) was initiated in 1994 and carried out during the SSHAC deliberations. Overlap in personnel between the YMP PSHA and the SSHAC provided knowledge of draft SSHAC guidance to the YMP PSHA.

The YMP PSHA was a landmark expert study in several ways. For example, it entailed the compilation and dissemination of a large amount of data that had been gathered in the site vicinity over the preceding 15 years. These data include extensive regional and local geologic maps, paleoseismic trenching (over 50 trenches), geomorphic analyses, and seismicity data that comprise the local network data, geodetic data, and a variety of geophysical analyses. Eighteen experts grouped into six interdisciplinary teams conducted the seismic source characterization (SCC), and seven experts conducted ground motion characterization (GMC). As is now typical, the seismic hazard assessments included fault displacement hazard as well as vibratory ground motion hazard. The study included several workshops and field trips and, following the process used in the earlier PVHA, featured proponents of alternative viewpoints and encouraged them to debate. The various expert roles defined in the SSHAC were incorporated explicitly into the project, were assigned to and known by the PSHA participants, and were emphasized by the TFIs (these roles were emphasized by the issuance of hats labeled "evaluator," "proponent," and "TFI"). A four-person participatory peer review panel was explicitly instructed to provide continual feedback throughout the project on both technical and process issues. The notion of the "interface" between SSC and GMC issues was identified as an area of importance, and a common workshop was held to discuss overlapping technical issues. Finally, the issue of the documentation of the experts' assessments—including their technical bases—was thoroughly addressed and an extensive documentation effort was undertaken.

Conducted in the late 1990s and sponsored by the NRC, the SSHAC Trial Implementation Project (TIP) (Savy et al., 2002) was intended to provide a forum for evaluating the SSHAC concept of a Technical Facilitator/Integrator (TFI) facilitating interactions among a panel of experts. In particular, the goal of the project was to conduct actual SSC evaluations for hazard analyses at two nuclear plant sites in the CEUS and to assess how expert interactions can lead to agreement among expert interpretations and better quantification of true uncertainties. Studies up to that time (e.g., the Yucca Mountain PVHA and PSHA) encouraged interaction among the experts and open discussion of alternative viewpoints. However, in the end, the TIP study treated each individual expert (or expert team) as a separate entity providing a separate assessment. SSHAC guidelines do not require consensus; however, after interactions, areas of common agreement among expert interpretations are usually reached. The TIP study focused on using the expert interaction process to debate the technical issues and to have the experts work as a team to look for areas of agreement as well as true disagreement. In a sense, this was an effort to construct a "composite" SSC model similar to the composite ground motion models used on some projects and discussed in the SSHAC document (NUREG/CR-6372, Section 5.7.3). The composite model would include the team's representation of knowledge and uncertainties. Likewise, an attempt was made to determine whether the areas of disagreement were, in fact, significant to the seismic hazard at the two sites. If not, the differences may not need to be propagated into the hazard calculations. Viewed now in light of the experience gained from more recent studies, the TIP process was very similar to the TI Team interaction process that is used for SSHAC Level 3 projects.

The first SSHAC Level 4 hazard analysis study conducted outside of the United States was the PEGASOS study conducted for four nuclear power plant sites in Switzerland (NAGRA, 2004; Abrahamson et al., 2002; Coppersmith et al., 2009). This also was the first commercial study conducted following the issuance of the SSHAC guidance for facilities under regulatory review[4]. In many ways, the study illustrated the value of a SSHAC process in dealing with contentious issues as well as the difficulties that arise from an independent third party attempting to understand and apply the guidance to a real project. Prior to the start of the study, the project management[5] gained approval for the SSHAC approach from the nuclear regulator HSK (now ENSI). In a rather unusual approach, the regulator HSK contracted the Participatory Peer Review Panel (PPRP), and reports were made to HSK rather than the project manager. As a result, the regulator HSK was kept abreast of the progress of the project. However, because the PPRP was not reporting to the project management and, in turn, to the sponsors of the study (i.e., the four nuclear utilities headed by Swissnuclear), the sponsors did not actively interface with the project via the PPRP and monitor its progress. As a result, the sponsors did not have a working knowledge of how the SSHAC process was being implemented during the course of the project.

The PEGASOS project was conducted successfully, accepted by the regulator HSK, and commended for use in subsequent risk analyses. The project provided major contributions in many technical and process-related areas including:

- Development and dissemination of digital databases to internationally distributed participants.

- Complete reexamination of historical seismicity and development of uniform earthquake catalogue.

- Highly interactive and informative workshops designed to juxtapose proponents of alternative data, models, and methods.

- Interface workshops to share SSC and GMC models and their overlap and implications.

- Numerous technical and process developments and innovations leading to the publication of over 20 papers in the professional literature.

- A major effort to develop feedback of various types to provide insight to experts prior to finalization of their models.

---

[4] Note that the TIP study was conducted by the NRC for purposes of research.
[5] The project was managed by NAGRA, the National Co-operative for the Disposal of Radioactive Waste.

- Development of the concept of a Hazard Input Document (HID) to summarize the technical assessments and transmit that information to the hazard analysts for calculation (see discussion in Section 4.7.2).

- Extensive documentation of both process and technical assessments.

The PEGASOS project also provided a number of lessons learned and opportunities for improvement. The expert assessments consisted of three panels: SSC, GMC, and site response. In retrospect, the site-response panel was placed in a difficult position as a result of being tasked with addressing both the uncertainties in the dynamic response of the near-surface materials beneath the plant sites and the uncertainties in the properties of the materials. It is likely that committing more resources to the gathering of site data to characterize site material properties would be more effective than addressing the issue using expert judgment. A lesson learned is that expert judgment should not be used as a substitute for reasonably obtainable data (discussed in Section 3.4). Another lesson learned is that the sponsors of the study should participate as observers at the various project workshops. This would ensure that the sponsors will be informed regarding the SSHAC process being conducted and the technical bases for hazard results when they became available.

At the conclusion of the PEGASOS project, simple comparisons were made between the hazard results from studies conducted some 15 years earlier and those from the PEGASOS study. Because the calculated mean hazard at the sites had gone up from the previous studies, NAGRA conducted a detailed analysis of the previous studies that clearly showed most of the difference was due to an inappropriate treatment of the ground motion aleatory variability in the previous studies and an error made in the previous calculations (NAGRA, 2004, Section 8.4.2). Such problems also have been found in comparisons between older hazard studies and modern PSHAs conducted at other sites (Bommer and Abrahamson, 2006). Nevertheless, the increase in calculated hazard results led the sponsor to conclude that the SSHAC process leads to "unconstrained accumulation of uncertainties" (Klügel, 2005). This conclusion was subsequently refuted (Musson et al., 2005). Swissnuclear has instituted a PEGASOS Refinement Project (PRP) that continues to use the SSHAC methodology and is designed to refine the original hazard results based on new site characterization data, an updated earthquake catalogue, and updated ground motion models.

Beginning in the early 2000s, interest grew on the part of the nuclear utilities in the United States to pursue Early Site Permits and Combined Operating License Applications for new nuclear power plants. Per regulatory guidance given in Regulatory Guides 1.165[6] (USNRC, 1997) and 1.208 (USNRC, 2007), the EPRI-SOG or LLNL PSHAs could be used as a starting point for the assessment of seismic hazard provided that they are updated to account for new knowledge gained since the conclusion of those studies many years earlier. In response to this need for an update of the ground-motion models, EPRI conducted a SSHAC Level 3 ground motion characterization study (2004) for central and eastern North America over a 1-year period. This study was the first avowed application of a SSHAC Level 3 process. A Technical Integrator (TI) was responsible for the assessments, and a panel of resource experts/proponents provided its views of the existing ground motion models and their applicability to the CEUS. The TI essentially used the panel as a "sounding board" in the development of a composite ground motion model that would meet the requirement of capturing the views of the informed technical community.

---

[6] RG1.165 has since been withdrawn.

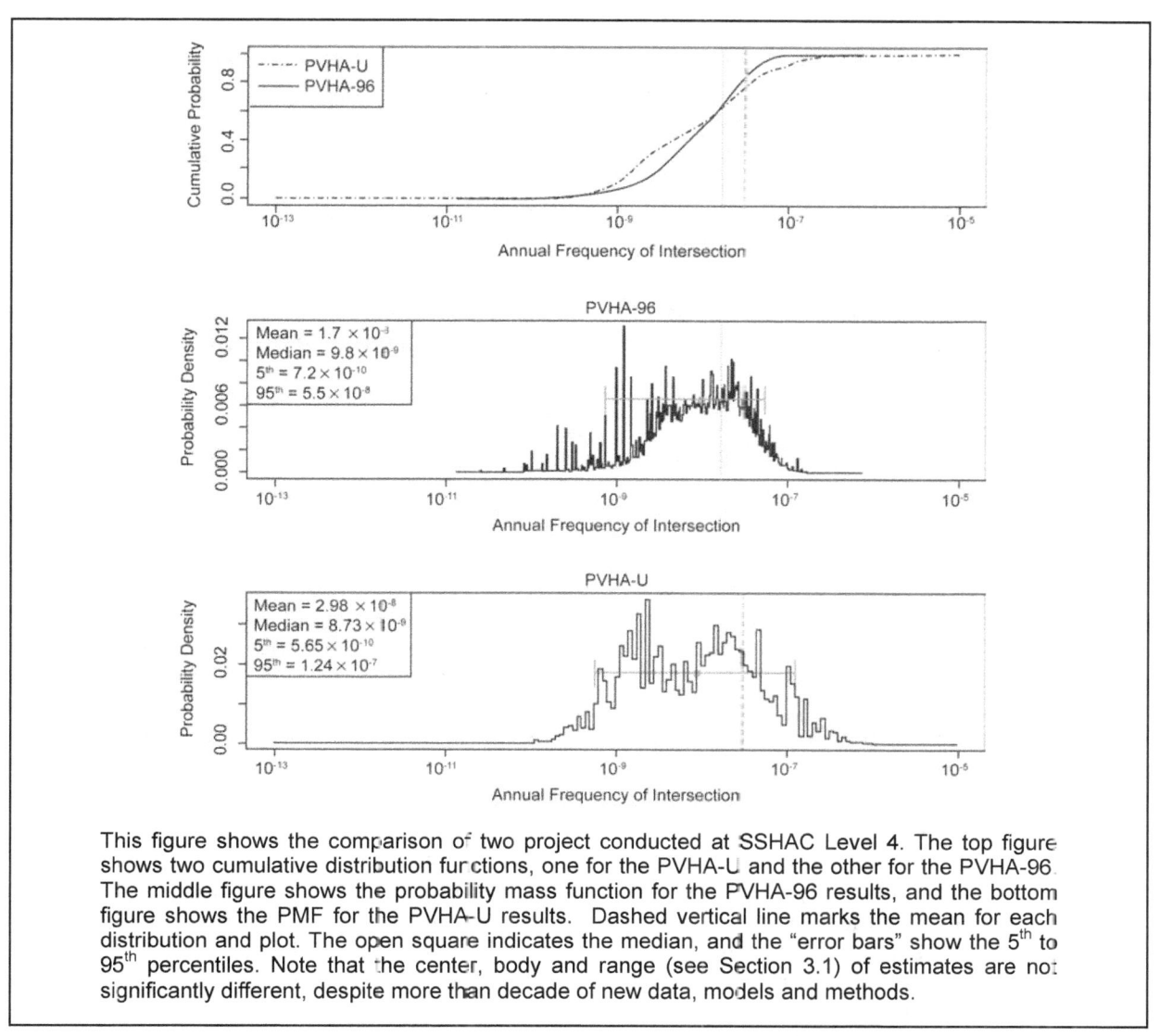

This figure shows the comparison of two project conducted at SSHAC Level 4. The top figure shows two cumulative distribution functions, one for the PVHA-U and the other for the PVHA-96. The middle figure shows the probability mass function for the PVHA-96 results, and the bottom figure shows the PMF for the PVHA-U results. Dashed vertical line marks the mean for each distribution and plot. The open square indicates the median, and the "error bars" show the 5th to 95th percentiles. Note that the center, body and range (see Section 3.1) of estimates are not significantly different, despite more than decade of new data, models and methods.

Figure 2-1.   Comparison of the Calculated Hazard Results from the PVHA (CRWMS M&O 1996) and the PVHA-U (SNL, 2008).

A lesson learned in the study was that technical experts would assume and maintain their roles as proponents without the TI naming them as part of a TI team and requiring them to take ownership of the resulting composite model.   Moreover, despite the value of having the proponent experts present their models, the TI has a distinctly different role that involves evaluating the views of the larger technical community. As such, a larger team of evaluators is needed. Subsequent Level 3 studies would expand the size of the TI team to include a larger number of experts and would require that all members of the team consider the views of the larger technical community and claim joint ownership of the results of the study (see discussion of ownership in Section 3.2).

Conducted over a period of more than 3 years, the Probabilistic Volcanic Hazard Analysis Update (PVHA-U) (SNL, 2008) took advantage of the SSHAC guidance for Level 4 studies and the experience gained from studies such as PEGASOS. The study was motivated by concerns from the NRC that new data developed for the Yucca Mountain Project could be interpreted such that it would lead to significant differences in the hazard results from that developed in the original PVHA a decade earlier. Despite developing sensitivity analyses that indicated the new data would not lead to a significant change in the hazard, the DOE decided in negotiation with the NRC to conduct an "update" to the previous PVHA. Because of the development of extensive new data in the region and significant advances in the methodologies for evaluating volcanic hazard, DOE and NRC decided that the "update" would not begin with the previous PVHA but would entail a completely new hazard analysis. Six of the expert panel members were part of the original PVHA panel, and two new members were added (two members of the PVHA panel had passed away in the intervening years). This was the first repeat of a Level 4 hazard study at the same site.

Several valuable aspects of the PVHA-U serve as lessons learned for future projects. As would be expected, the recent advances in the development, archiving, and dissemination of digital databases were used in developing the project database. Considerable effort in the form of quantitative analyses and consideration of analogous studies was made early in the project to identify the important technical issues that would drive the hazard results. This information, in turn, provided an objective means of focusing the subsequent effort by the experts. Early in the project, the expert panel identified a series of important data collection activities, including drilling several boreholes to investigate shallow aeromagnetic anomalies, that would serve to significantly reduce its uncertainties in key inputs to the analysis. DOE conducted these focused data collection activities in a timely fashion, and the experts duly evaluated the results in the development of their models. Because the experts relied largely on their experience at analogue localities, they found it useful to take field trips to observe the analogue information on a first-hand basis. An emphasis was placed on providing timely and useful feedback to the experts on their preliminary interpretations, prior to finalization. The feedback was designed to show which issues were most important to the hazard results as well as the most significant contributors to the uncertainty in the hazard results. Importantly, the center, body, and range of the hazard results between the two studies were very similar (Figure 2-1) (SNL, 2008, section 4.3) despite the passage of 12 years since the original PVHA, collection of extensive new site-specific and regional data, and advances in the characterization tools for volcanic hazard. Subsequent risk analyses showed that the differences between the hazard results from the two studies do not lead to significant differences in the mean risk (DOE, 2008, p.5.4-6).

Much of the engineering community in the nuclear field relies on codes and standards that have been developed as consensus standards of practice by the larger community. In the mid-2000s, working groups of American Nuclear Society Standards Committees developed a series of standards dealing with probabilistic seismic hazard analysis. The Standard ANSI/ANS-2.29-2008 *Probabilistic Seismic Hazard Analysis* is significant because it refers specifically to the SSHAC guidance and outlines a process for conducting a PSHA that is consistent with the SSHAC methodology. The document defines the roles of various participants in a manner that reflects the key roles of a SSHAC process. Significantly, the document also identifies the key differences in the four SSHAC Levels (termed "PSHA Levels" in the ANSI/ANS standard), and it discusses the issue of selecting an appropriate Level for a particular application (see discussion in Section 4.2). Three criteria are identified as useful for informing the decision on PSHA Level: (1) the risk significance of the facility (expressed in ANSI/ANS-2.26-2004 as the Seismic Design Category from ASCE-43-05), (2) the nominal ground motion hazard level (from the U.S. National Seismic Hazard Maps), and (3) the level of uncertainty and scientific controversy in the tectonic regime.

At the time of this writing, four SSHAC Level 3 studies are being conducted that provide a wealth of experience regarding the approaches that can be successfully implemented. The Central and Eastern United States Seismic Source Characterization (CEUS SSC) project was unique in its sponsorship and the participation by the three key stakeholders for nuclear facilities in the United States—the NRC, DOE, and nuclear utilities (represented by EPRI). By establishing a "community-based" approach that includes broad sponsorship, the ownership of the resulting SSC model will be broad-based across all parties. This will eliminate the dichotomies that came from the EPRI-SOG and LLNL studies. The Next Generation Attenuation Relationship for Eastern North America (NGA-East) project, now underway, is following a similar model and has similar sponsorship. The study is the GMC counterpart to the CEUS SSC study, and together they will fully replace the EPRI-SOG and LLNL studies and the EPRI ground-motion models. The Canadian electric utility BC Hydro is nearing completion of a SSHAC Level 3 PSHA for its large service area in western Canada. The project has met significant challenges in the requirement to characterize the seismic sources and appropriate ground-motion models for a wide range of tectonic environments ranging from subduction zone sources, to active crustal faults, to stable continental regions. This is the first nonnuclear application of the SSHAC Level 3 methodology, and the results will be used in system-wide risk analyses. A PSHA is being conducted for the licensing of a new nuclear power plant at the Thyspunt site in South Africa using a SSHAC Level 3 study. The Thyspunt project is the first SSHAC Level 3 site-specific PSHA for a new nuclear power plant site and also the first application of the higher SSHAC Levels in the developing world.

2.4 The Need for Implementation Guidance

A review of the history of expert assessment studies and the development of associated guidance documents (as described above) indicates the following points: (1) the existing SSHAC guidance effectively defined basic concepts that have proven to be useful in actual application, (2) sufficient actual project experience with SSHAC Level 3 and 4 studies exists to be able to benefit from lessons learned, and (3) a need exists to develop new implementation guidance that captures what works well and avoids the pitfalls. Expert assessment methodologies generally—and SSHAC implementation processes specifically—are evolutionary and continually improve with each project application as a result of innovation and lessons learned. For example, studies have shown that the makeup of the PPRP in Level 3 and 4 studies can have a profound effect on the course of a study and on the credibility of the study results to regulators and outside observers. Another example is that the timing and content of feedback can be very important to ensuring that all of the experts understand both the implications of their preliminary assessments and where to focus their efforts in the finalization of their assessments.

Table 2-2 provides representative examples of multi-expert hazard studies and associated guidance. Because of the experience gained since the publication of the SSHAC report NUREG/CR-6372, it is appropriate to assess and capture the lessons learned and to provide additional guidance on the specific project steps that have been shown to be effective in attaining the SSHAC goals. At the same time, there is no presumption that documenting today's best advice regarding SSHAC implementation guidance should put an end to the evolutionary process that has characterized the past 20 years. On the contrary, this information and guidance is provided with the assumption—and the belief—that further advances will continue to be made in the future.

Table 2-2.
Representative Examples of Multi-Expert Hazard Studies and Associated Guidance

Study	Year	Discussion of Significance
WASH-1400 Reactor Safety Study	1975	The first U.S. probabilistic analysis of nuclear power plant safety was the WASH-1400 study (USNRC, 1975). The AEC/NRC staff performed the study under the direction of consultant Norman C. Rasmussen of MIT over a period of 3 years with the participation of about 60 staff and consultants. The study is considered the first application of the probabilistic risk analysis (PRA) methodology to nuclear power plants and resulted in quantitative estimates of the accident risk to the public. Despite the participation of a large contingent of the technical community in the study, formal methods of eliciting expert judgment and quantifying uncertainties were not used. Subsequent peer review of the study generally endorsed the PRA methodology but suggested that the uncertainties associated with many of the key inputs were quite significant. The study was later replaced by the NUREG-1150 (USNRC, 1991a) study discussed below. Note that the WASH-1400 study dealt with *risk* and not with *hazard*.
NUREG-1150 (USNRC 1991) Severe Accident Risks: An Assessment for Five U.S. Nuclear Power Plants	1987 -1991	Beginning in the mid-1980s, a number of studies incorporated the use of formal expert elicitation as a means of capturing the knowledge and uncertainties needed to characterize the key inputs to probabilistic hazard and risk analyses. NUREG-1150 (USNRC, 1991a) documents a large study designed to estimate the uncertainties and consequences of severe core damage accidents in selected nuclear power plants. Seven panels of experts were involved in an extensive expert elicitation process (USNRC, 1991a), and PRAs were conducted for each of five nuclear power plants having different plant and containment designs. Note that the NUREG-1150 study actually dealt with *risk* not with *hazard*.
EPRI-SOG and LLNL PSHA Studies	1988, 1989	At about the same time as the NUREG-1150 study (USNRC, 1991a), two large probabilistic seismic hazard studies were conducted to assess the hazard at commercial power plant sites in the central and eastern United States—the Electric Power Research Institute's Seismicity Owner's Group (EPRI-SOG) study (EPRI-SOG 1988, 1989) and the Lawrence Livermore National Laboratory (LLNL) study (Bernreuter et al. 1989). The uncertainty treatment in the LLNL study was updated in Savy et al. (1993). Both studies used multiple experts to capture the uncertainties associated with earthquake hazards, although different approaches were used in the expert elicitation process. For example, in the EPRI-SOG study, multiple workshops were held to discuss technical issues, but the LLNL study did not include workshops or other interactions among the experts. The LLNL study used an outside expert panel for ground motion estimation while the EPRI-SOG study left this assessment to the project team. The LLNL study quantified uncertainties based on Monte-Carlo simulation of continuous parameters while the EPRI-SOG used logic-trees to quantify uncertainties.
EPRI-SOG and LLNL PSHA Studies	1988, 1989	The EPRI-SOG project, in which seismic hazard was assessed for 69 nuclear power plant sites in the central and eastern United States, was important for developing methodologies and procedures on the conduct of a formal expert assessment utilizing multiple experts and was a direct predecessor to the SSHAC study. This hazard analysis focused on developing a methodology for PSHA that included a structured process for interpreting the tectonics of an area to define the seismic source zones and used statistical analyses of a historical earthquake catalog to develop earthquake size and rate parameters. An initial part of the study involved compilations of comprehensive geophysical and seismological databases. These databases were then distributed to six earth science teams composed of individuals representing the fields of seismology, geology, and geophysics. The teams independently developed seismic source zones and associated seismicity parameters for the area of focus, explicitly accounting for uncertainties in the evaluations using alternative, weighted interpretations for individual zones or features. To implement the methodology, numerous workshops and meetings with project participants were convened, and the methodology team worked with the participants to assess their scientific judgments and to format those judgments to be suitable for hazard calculations.

Table 2-2. (Continued)
Representative Examples of Multi-Expert Hazard Studies and Associated Guidance

Study	Year	Discussion of Significance
Satsop PSHA	1990	Considerable uncertainty exists regarding the earthquake potential of the Cascadia subduction zone in the Pacific Northwest of the United States due to the historically aseismic nature of the interface between the Juan de Fuca and North American plates. A PSHA involving formal assessment of multiple experts was conducted in this region for the proposed Satsop nuclear power plant site in western Washington State (Coppersmith and Youngs, 1990). To develop a complete seismic source characterization spanning the range of interpretations regarding the earthquake potential of Cascadia, a group of 14 experts was selected based on their experience in the region and at convergent margins worldwide. These experts assessed source characteristics, including subduction zone geometry of the plate interface, intraslab, and crustal seismic sources; probability that each potential source is seismogenic; expected locations and dimensions of rupture; maximum earthquake magnitude; earthquake recurrence models; paleoseismic recurrence intervals; plate convergence rate; and seismic coupling. Important sources of uncertainty included alternative conceptual models regarding the nature and rate of seismogenic convergence across a plate interface that has not experienced moderate to large earthquakes during the historical period of about 100 yr. Prior to the elicitation, new geologic evidence of episodic coastal subsidence had become available, which was interpreted by some as evidence for prehistorical seismogenic coupling along the plate interface. The results of the seismic hazard analysis were submitted by the electric utility to the NRC as part of licensing activities.
WIPP Study	1991	A formal expert judgment elicitation process was followed to assess long-term radionuclide releases from the Waste Isolation Pilot Plant (WIPP), an underground radioactive waste repository in southeastern New Mexico (Trauth et al., 1991). A performance assessment was conducted by four experts, who also developed probability distributions for the concentration of dissolved radionuclides at the WIPP.
CNWRA Climate Study	1993	The Center for Nuclear Waste Regulatory Analysis conducted an expert elicitation to estimate the future climate in the Yucca Mountain Region (DeWispelare et al., 1993). The objectives of this study included acquiring expertise in the expert elicitation process to aid in the reviews of DOE's use of expert elicitation, investigating aggregation and consensus-building techniques for expert panels, and contributing to the development of NRC guidance on expert judgment elicitation. Future climate was selected for study because the uncertainties associated with this topic are large considering climate variance during the Quaternary period, and the climatic impact on groundwater infiltration can be important to repository performance. Five panel members participated in the study and developed probability distributions for a variety of issues. Each panel member prepared a position paper describing their judgments and the technical bases for the judgments.

Table 2-2. (Continued)
Representative Examples of Multi-Expert Hazard Studies and Associated Guidance

Study	Year	Discussion of Significance
EPRI Demonstration Project	1993	The Earthquakes and Tectonics Expert Judgment Elicitation Project was sponsored by the Electric Power Research Institute (Coppersmith et al., 1993). The objectives of this study were two-fold: (1) to demonstrate methods for the elicitation of expert judgment, and (2) to quantify the uncertainties associated with earthquake and tectonics issues for use in the EPRI-HLW (high level waste) performance assessment. Specifically, the technical issue considered was the probability of differential fault displacement through the proposed repository at Yucca Mountain, Nevada. For this study, a strategy for quantifying uncertainties was developed that relies on the judgments of multiple experts. A panel of seven geologists and seismologists was assembled to quantify the uncertainties associated with earthquake and tectonics issues for the performance assessment model. A series of technical workshops focusing on these issues was conducted. Finally, each expert was individually interviewed in order to elicit his judgment regarding the technical issues and to provide the technical basis for his assessment. This report summarizes the methodologies used to elicit the judgments of the earthquakes and tectonics experts and summarizes the technical assessments made by the expert panel.
PVHA Yucca Mountain	1996	A probabilistic volcanic hazard analysis (PVHA) was carried out for Yucca Mountain (CRWMS M&O, 1996; Kerr, 1996; Coppersmith et al., 2009) during a three-year period. The study was conducted using an expert elicitation process and was conducted at the same time that the NRC was developing regulatory guidance on expert elicitation (Kotra et al., 1996). Experts with the needed expertise in volcanic studies in the southern Great Basin or similar extensional environments were assembled to evaluate volcanic hazard in the Yucca Mountain region. The probability of a future volcanic event disrupting the repository at Yucca Mountain was the focus of this study. Ten experts were selected from a pool of 70 candidates; the results of 15 years of geologic and volcanic data-collection activities for the site were made available to them. Facilitated interaction of the experts was encouraged in multiple workshops and two field trips. The PVHA expert assessments focused on spatial models defining the future locations of volcanic events and temporal models defining the rate of occurrence of events. To make their assessments and to quantify the associated uncertainties, the experts considered the geologic data in the Yucca Mountain region as well as their own experience at analogue basaltic volcanic fields. The results of the hazard assessment provided inputs to the total system-performance assessment for the repository system for the 10,000-year compliance period and were made part of the License Application submitted to the NRC. This first PVHA was completed in 1996 and an update to the PVHA was completed in 2008 (discussed below).
NUREG-1563 (Kotra et al., 1996)	1996	In 1996, NRC developed regulatory guidance NUREG-1563 specifically for the high-level waste program (Kotra et al., 1996). The document identifies a number of conditions that warrant the use of expert elicitation, including "empirical data [that] are not reasonably obtainable; uncertainties [that] are large and significant to a demonstration of compliance; and more than one conceptual model can explain and be consistent with the available data." In that guidance, expert elicitation steps are identified, including defining objectives, selecting the experts, eliciting expert judgments and documenting the technical bases for the assessments. The guidance also discusses the benefits of expert interaction and discussion, and alternative approaches to combining or aggregating the judgments of multiple experts. Reference is made to the SSHAC study, which was underway.

Study	Year	Discussion of Significance
Yucca Mountain Viability Assessment	1996-98	As part of studies for Yucca Mountain, the DOE completed five expert elicitations to support the total system performance assessment for the viability assessment from 1996 to 1998. The purpose of these five expert elicitations was to quantify uncertainties associated with key models and to provide a perspective on modeling and data collection activities that could help to characterize and reduce uncertainties. These elicitations focused on five models: • Saturated zone flow and transport (CRWMS M&O, 1998c) • Unsaturated zone flow (CRWMS M&O, 1997) • Near-field environment and altered zone coupled effects (CRWMS M&C, 1998a) • Waste form degradation and radionuclide mobilization (CRWMS M&C, 1998d) • Waste package degradation (CRWMS M&O, 1998e). The meetings of the expert panel were structured, facilitated interactions in workshops and, for some projects, field trips. A team consisting of technical and normative experts facilitated the workshops. The workshops were designed to identify the significant issues, available data, alternative models, and uncertainties related to each process model. The expert panel members were given detailed summaries and presentations of available data and models and the status of various components of the modeling and testing program. Debate and technical challenge of alternative interpretations was encouraged to ensure that uncertainties were identified. Researchers from a variety of organizations participated as resource experts, including national laboratories, U.S. Geological Survey, universities, and consulting companies. Resource experts presented pertinent data sets and proponent experts presented alternative models and methods.
SSHAC	1997	In 1997, after four years of deliberations, the Senior Seismic Hazard Analysis Committee released its report as NUREG/CR–6372 (summarized in Budnitz et al., 2006). The report addresses why and how multiple expert judgments–and the intrinsic uncertainties that attend them–should be used in PSHA for critical facilities such as commercial nuclear power plants. More specifically, the SSHAC Guidelines are concerned with how to capture, quantify, and communicate the uncertainties expressed by multiple experts. SSHAC was originally convened to review and understand the differing PSHA results obtained by the EPR-SOG (1988) and the LLNL (Bernreuter et al., 1989) for the same nuclear facilities in the eastern United States. According to NUREG/CR-6372, the differing PSHA results stemmed from the different ways that the two studies elicited and aggregated the differing interpretations, judgments, and models of their experts. After considering the issue, NUREG/CR-6372 proposed a process for obtaining and aggregating expert interpretations, judgments, and models that is quite different from those used in conventional elicitation/aggregation procedures (see references in Appendix J of the SSHAC Guidelines). This process begins with diverse inputs, such as differing models and interpretations obtained from multiple experts, which are then evaluated through an interactive process overseen by a technical integrator (TI) or technical facilitator/integrator (TFI). The SSHAC guidance provides advice for four "study levels," which are differentiated as a function of the importance, complexity, diversity of views and contentiousness of an issue. The level of study required for a hazard analysis is related to factors such as the regulatory framework, the resources (money and time) available to conduct the study, perceptions of the importance of the project, and scheduling constraints. Regardless of the level of study, the goal is the same: to provide a representation of the center, body, and range of the views of the informed scientific community regarding the important components and issues, and, finally, the hazard.

Table 2-2. (Continued)
Representative Examples of Multi-Expert Hazard Studies and Associated Guidance

Study	Year	Discussion of Significance
Yucca Mountain PSHA	1998	DOE initiated a four-year probabilistic seismic hazard analysis (PSHA) for Yucca Mountain using an expert elicitation process(CRWMS M&O, 1998b). The methodology followed was consistent with SSHAC recommendations for a Level 4 PSHA as well as guidelines established by the NRC for expert elicitations (Kotra, et al., 1996). Experts with the needed range of expertise in regional and local earthquake and fault tectonics, earthquake physics, ground motion modeling, and seismic hazard analyses were assembled to evaluate seismic hazards in the Yucca Mountain region. They characterized and assessed the uncertainty of seismic sources, earthquake recurrence, ground motion models, and fault displacement models for faulting conditions known to be present in the vicinity of the Yucca Mountain site. Six SSC teams, with expertise in Basin and Range Province earthquake tectonics, earthquake seismology, and Quaternary fault displacement modeling, evaluated seismic source, fault displacement, and associated uncertainties using a structured elicitation process were assembled. A separate panel of seven ground motion experts was convened using a similarly structured elicitation process to evaluate ground motion attenuation and associated uncertainties. Geologic and seismologic studies of the Yucca Mountain region had been conducted during the preceding 15 years, and the data developed from these studies provided a fundamental resource for the SSC experts. Field trips were held with the SSC experts to provide them with opportunities to observe the field relationships on which interpretations of the paleoseismic behavior of faults were based. The study included multiple workshops designed to facilitate the interactions among the experts and to assist them in their evaluations. GMC at Yucca Mountain included evaluations of empirical ground motions recorded worldwide and at Yucca Mountain, region-specific numerical simulations for the Yucca Mountain sources and crustal structure, and ground motions from nuclear explosions at the adjacent Nevada Test Site. Using seismic source and ground motion inputs, the seismic hazard was calculated and expressed as a probability distribution on the annual frequency at which levels of ground motion or fault displacement will be exceeded. These results form part of the bases for developing pre-closure seismic design inputs and provide information on the frequency of occurrence of potentially disruptive ground motions for assessment of long-term performance of the repository (CRWMS M&O 1998, Chapter 1). The Yucca Mountain PSHA placed an emphasis on capturing uncertainties in paleoseismic data and expressing probabilistic fault displacement hazard (Stepp et al., 2001; Youngs et al., 2003). The results of the study were incorporated into the License Application submitted to the NRC for the repository at Yucca Mountain.

Study	Year	Discussion of Significance
SSHAC Trial Implementation Project	2002	The scope of Trial Implementation Project (Savy et al. 2002) was designed to test the recommendation of the SSHAC on the characterization of the seismic sources, and to finalize the development of ground motion attenuation models for eastern North America started by SSHAC. The study had the goal of testing and implementing the SSHAC guidelines for the specific case of the southeastern United States and of two nuclear plant sites in that region, namely Vogtle and Watts Bar. Workshops and expert elicitations were held in accordance with SSHAC principles, with emphasis on seismic source characterization. This project showed that the TFI procedures can lead to an unusual degree of agreement among experts through thorough discussion of the available data, and through interaction amongst the experts. Together with the focusing effect of the TFI, this leads to narrower margins of variation without any coercion. For the southeastern U.S. this led to an integrated map of source zones that incorporated the opinions of all the experts involved, even though they began with fairly different source zone maps. The main purpose of this process was to minimize the unnecessary, or artificial, diversity by making sure that those interpretations that appeared different, were indeed different. Those that were not were folded into a common interpretation, with some uncertainty.
PEGASOS (Switzerland) PSHA and PEGASOS Refinement Project	2004 2011	Although Switzerland is generally considered to have a low to moderate level of seismicity, the Swiss Federal Nuclear Safety Inspectorate (HSK, subsequently changed to ENSI) identified seismic hazard as a potentially significant contributor to the risk at four nuclear power plant sites (Mühleberg, Gösgen, Beznau, and Leibstadt). The HSK also identified the need to update the seismic hazard analyses at the sites and requested that the Swiss electric utilities conduct a PSHA following SSHAC Level 4 methodologies. Under the direction of National Cooperative for the Disposal of Radioactive Waste (NAGRA), a PSHA was conducted for Swiss nuclear power plant sites. The study has since become known under the name of the "PEGASOS Project" (Probabilistische Erdbeben-Gefährdungs-Analyse für KKW-StandOrte in der Schweiz) (NAGRA, 2004; Abrahamson et al., 2002; Coppersmith et al. 2009). The objective of the project was to assess the relevant earthquake induced ground motions at the building foundation levels of the four sites, which would be used subsequently for probabilistic safety analyses. A full-scope formal expert assessment process was used, including dissemination of a comprehensive database, multiple workshops for identification and discussion of alternative models and interpretations, assessment interviews, feedback to provide the experts with the implications of their preliminary assessments, and full documentation of the assessments. The study brought together experts from all over Europe. Four teams, consisting of three experts each, conducted the seismic source characterization, five individual experts addressed ground motion characterization, and four experts characterized the site response. The entire study was subject to participatory peer review by an HSK Review Team, which monitored and provided feedback on the procedural and technical aspects of the project, as well as provided a review of the final report. Because of the large uncertainties associated with the site response and, to a lesser extent, the ground motion aspects of the study, a PEGASOS Refinement Project was begun in 2008 with the aim of incorporating new data regarding the site conditions at the four nuclear power plant sites and incorporating new ground motion models. The implications of an updated earthquake catalog is also being incorporated into the PEGASOS Refinement Project.

Table 2-2. (Continued)
Representative Examples of Multi-Expert Hazard Studies and Associated Guidance

Study	Year	Discussion of Significance
EPRI Ground Motion Study	2004, 2006	A SSHAC Level 3 ground motion characterization study for central and eastern North America was conducted by EPRI (2004) during a one-year period and was the first avowed application of a Level 3 process. The development of a composite understanding of ground motion attenuation is a contentious, complex issue, and the results are significant to the resultant hazard. Thus a SSHAC-defined Study Level 3 analysis was utilized. A single Technical Integrator (TI) was responsible for the development of the composite distribution on ground motion based on evaluations of available information, including interactions with ground motion experts. The TI brought together a panel of ground motion experts, including proponents of available models, for a series of three workshops for debate and interaction. The TI interacted with the Expert Panel to understand the features of available models and develop a basis for model evaluation and an approach for estimating the scientific community distribution. The product of the study was a ground motion attenuation model defined by a set of equations and coefficients for estimating ground motion measures and their aleatory variability as a function of earthquake magnitude and source-to-site distance. The model includes the epistemic uncertainty in the median estimate of ground motions and in the aleatory variability. The model is applicable to two general regions in the central and eastern United States: Mid-Continent (CEUS excluding Gulf Coast) and the Gulf Coast. The model is applicable to three classes of seismic sources: general conditions involving area sources; distant, active, large magnitude sources; and nearby large magnitude faults. Shortly after completion of the EPRI (2004) study, a SSHAC Level 2 study was conducted that initially was intended to deal with upper truncation of the ground motion residual distribution (Strasser et al., 2008), and later focused on the value of the standard deviation for the ground motion variability for the central and eastern United States (CEUS). The value of the standard deviation in the models developed in the EPRI (2004) ground motion study was much larger than recent studies of large data sets of ground motions applicable to the WUS have shown. An evaluation of differences in the standard deviation in the CEUS and WUS based on the variability of the source, path, and site terms indicated that the WUS intra-event standard deviations are generally applicable to the CEUS with some epistemic uncertainty in the effect of focal depth at short distances and that the inter-event standard deviations may be larger in the CEUS than in the WUS based on larger variability in the stress-drops. Alternative models for the total standard deviation (combined intra-event and inter-event) were developed that can be applied to the CEUS. Overall, these new models show a significant reduction in the standard deviation, particularly at short distances (EPRI, 2006). This lower value of the standard deviation tends to reduce the computed hazard as compared to the EPRI (2004) models.

Table 2-2. (Continued)
Representative Examples of Multi-Expert Hazard Studies and Associated Guidance

Study	Year	Discussion of Significance
PVHA-Update Yucca Mountain	2008	In 2005–2007, DOE conducted the probabilistic volcanic hazard analysis update (PVHA-U) involving a panel of eight experts (six of whom were on the original PVHA expert panel) and a SSHAC Level 4 process (Sandia National Laboratories 2008). This first-of-a-kind update of a major formal expert assessment was motivated by the availability of new data subsequent to the 1996 study and by advances in approaches for modeling volcanic hazard. Newly available data included new high-resolution aeromagnetic data and drilling/geochronology/geochemical data. Several aeromagnetic anomalies were identified in alluvial basins in the region that were postulated to represent buried basaltic bodies. The drilling data provided information on the composition, depth, and age of the igneous features giving rise to the anomalies. At their request, the experts also were provided with information regarding characteristics of analogue volcanic events in the region. Conceptual models developed by the experts for the future spatial distribution of volcanic events included spatial smoothing of applicable past events, consideration of parametric field shapes, and identification of alternative tectonic zones defined by variations in the rate of volcanism. Temporal models developed by the experts included models of recurrence rate as a Poisson process defined by observed events of specified age, temporally clustered models, and time-volume models that account for the decline in eruptive volumes over the past several million years. In the PVHA-U, particular emphasis was placed on defining the characteristics of future volcanic events including the number and dimensions of dikes and eruptive conduits in an event, the geometry of these features, and the expected eruptive type. The experts made assessments of the volcanic hazard over future time periods of 10,000 years and 1,000,000 years. The results of the PVHA-U were submitted to the NRC as part of the License Application for the Yucca Mountain repository.
ANSI/ANS Standard-2.29-2008 PSHA	2008	Over a period of several years, ANSI/ANS-2.29-2008 (2008) was developed with the purpose of providing criteria and guidance for performing a probabilistic seismic hazard analysis for the design and construction of nuclear facilities. The criteria provided in the Standard address various aspects of conducting PSHAs including selection of the process or methodology and the level of seismic hazard analysis appropriate for a given seismic design category, seismic source characterization, ground motion estimation, site response assessment, assessment of aleatory and epistemic uncertainties, and PSHA documentation requirements. The "PSHA Levels" defined and described in the Standard are essentially the same as the SSHAC Study Levels 1-4, and further descriptive discussion is provided to assist the practitioner in making a selection of the appropriate level. Other recommendations related to expert roles, expert interactions (workshops), peer review, and documentation are all quite comparable to, and supportive of, the recommendations made in NUREG/CR-6372.

Study	Year	Discussion of Significance
CEUS SSC Project	(completed 2011)	The Central and Eastern U.S. Seismic Source Characterization for Nuclear Facilities (CEUS SSC) Project—jointly sponsored by NRC, EPRI, and DOE—is aimed at developing a comprehensive seismic-source model for the entire CEUS (EPRI, 2008). This SSHAC Level 3 study began in September 2008 and finished in December 2011, and will supersede the 1986 EPRI-SOG and LLNL seismic source models. The goal of the CEUS SSC Project is to develop a stable and long-lived CEUS SSC that includes (1) full assessment and incorporation of uncertainties, (2) the range of diverse technical interpretation, (3) consideration of an up-to-date database, (4) proper and appropriate documentation, and (5) peer review. The CEUS SSC project team consists of program and project management, a TI Lead; TI team (consisting of about 15 seismologists, geologists, and hazard analysts); a participatory peer review panel (PPRP); specialty contractors; sponsors; and funding agency experts. The study includes the development of a CEUS geological, geophysical, and seismological database in geographic information system (GIS) format and an updated earthquake catalog that merges and reconciles several regional catalogs and develops uniform moment magnitudes. The study is centered around three workshops that (1) identify hazard-significant SSC issues and identify and discuss important databases; (2) present, discuss, and debate alternative interpretations of significant SSC issues with proponents of alternative models; and (3) present the preliminary SSC model and discuss hazard feedback and sensitivity analyses. Following the second workshop, a preliminary SSC model was developed that was the basis for hazard and sensitivity calculations discussed at the third workshop. Following the third workshop, a draft SSC model was developed with all uncertainties, and the draft CEUS SSC project report was developed for review. Following review of the draft SSC model, the final SSC model was developed and documented in the final project report. A number of working meetings of the TI team were held throughout the project to develop the SSC model. The eight-member PPRP is responsible for review of both the technical and process aspects of the project.

Table 2-2. (Continued)
Representative Examples of Multi-Expert Hazard Studies and Associated Guidance

Study	Year	Discussion of Significance
BC Hydro PSHA	(expected completion 2012)	Using a SSHAC Level 3 process, BC Hydro is conducting a PSHA to provide an up-to-date and comprehensive assessment of the seismic hazard in British Columbia (BC) and a basis for estimating ground motions at about 46 project sites throughout the region. BC Hydro's desire is to develop a PSHA model that will provide a sound, stable measure of ground motion hazards in the province for 10 to 15 years. The study region covers the majority of BC, part of the northwestern United States, and part of Alberta to the east. The region is tectonically diverse with seismic sources ranging from plate boundary sources, active faults in the coastal region, to a stable continental region in the eastern part of the study area. The seismic source characterization (SSC) and ground motion (GMC) components of the study both consist of TI teams and Evaluation Staff that are responsible for carrying out the technical work. The SSC Evaluation Staff has subdivided the source characterization effort according to seismic source type, but it is expected that all members of the TI team and Evaluator Staff will assume ownership of all aspects of the model. An innovative feature of the GMC modeling approach in this project was the development of a single new subduction GMPE—following evaluation of existing equations as being inadequate—and then evaluating epistemic uncertainty and adding it to this single model to create the logic-tree branches. Multiple workshops are being held to ensure that the views of the larger technical community are being identified and made part of the evaluation. The PPRP plays an important role in ongoing review of the process and technical evaluations being made. The technical evaluations are tackling a number of major technical issues of importance to seismic hazard including characterizing subduction zone sources using a combination of geologic, seismologic, and geophysical datasets; evaluating the use of geodetic data to define earthquake recurrence; defining seismic sources in regions with very little geologic data; and updating and customizing ground motion models for applicability to the BC region.
Thyspunt, South Africa PSHA	(expected completion 2013)	A site-specific SSHAC Level 3 PSHA study is being conducted for purposes of licensing a nuclear power plant on the coast of South Africa. A Level 3 study was selected as the appropriate Level in order to provide significant levels of regulatory assurance for purposes of licensing. The study takes advantage of many years of geologic, geophysical, and seismological data collection in the site vicinity and the region by the Council for Geosciences. In addition, a program of focused data collection was indentified and prioritized to provide reductions in the uncertainties associated with key technical issues for the PSHA. The project is divided into TI Teams for seismic source characterization and for ground motion characterization. The Thyspunt PSHA is decidedly an international study with participants on the TI Teams; resource experts and PPRP representing South Africa, several countries in Europe, and the United States. The PPRP consists of six members with extensive experience in SSHAC processes as well as PSHAs worldwide. The National Nuclear Regulator currently has not issued regulatory guidance regarding acceptable methods for conducting a PSHA, but the existing regulations call for conducting probabilistic risk analyses and the project is using the SSHAC methodology because of its acceptance by the NRC and other regulatory groups in the United States. The study includes the workshops, expert interactions, and documentation that characterize a complete SSHAC Level 3 process.

Table 2-2. (Continued)
Representative Examples of Multi-Expert Hazard Studies and Associated Guidance

Study	Year	Discussion of Significance
NGA-East Project	(expected completion 2014)	With support from the NRC, EPRI and DOE, and the USGS, the Pacific Earthquake Engineering Research Center (PEER) is coordinating a comprehensive multidisciplinary program to develop Next Generation Attenuation Relationships for Central and Eastern North America (NGA-East). This follows on the very successful multi-institution, multi-investigator, multisponsor collaborative Next Generation Attenuation Relationship (NGA-West) program (originally referred to as NGA) that developed new ground motion prediction models for the western United States. Because of the high levels of uncertainty and the need for regulatory assurance, an NGA-East study was instituted with the objective to develop a new set of comprehensive and broadly accepted attenuation relationships for the Central and Eastern North America (CENA). To support the NGA-East attenuation modelers, several sets of supporting projects will be defined, initiated, and coordinated. These include development of a ground motion database, and supporting computer simulation studies. Unlike the NGA-West project, the NGA East study is being conducted using a SSHAC Level 3 process. This project is the GMC counterpart to the CEUS SSC project noted above. All of the principal participants have been identified and early work has been initiated to develop the project databases. The PPRP for this study includes international participants.

Future improvements may be made in finding ways—perhaps through establishing strict criteria—to increase the efficiency of SSHAC Level 3 and 4 studies such that they can be carried out in shorter periods of time or more cost effectively. In addition, rapid advances in technology have revolutionized our ability to compile, organize, and disseminate data. The results of this progress have found their way into SSHAC hazard studies and have led to dramatic advances in the data compilation step of SSHAC projects. Future studies most likely will continue to improve with the application of data management and interpretation technologies including perhaps real-time feedback regarding the implications of new data collection efforts on key uncertainties in the hazard model. Because new and better approaches will likely continue to evolve, the SSHAC implementation guidance provided in this document is intended to (1) strike a balance between providing guidance and being overly prescriptive and (2) define successful process steps without stifling innovation.

3. KEY SSHAC CONCEPTS

This Chapter presents an overview of the key concepts that define and distinguish a Senior Seismic Hazard Analysis Committee (SSHAC) process. Most of these concepts are already clearly defined in the original SSHAC guidelines (NUREG/CR-6372), but the explanations given here attempt to clarify some ambiguities in the original documentation and identify subtle changes that the experiences of implementation have suggested could be beneficial.

3.1 The Center, Body, and Range of the Informed Technical Community

The key statement in the SSHAC guidelines that encapsulates the ethos of the SSHAC approach is as follows: *"Regardless of the scale of the PSHA study the goal remains the same: to represent the center, the body, and the range that the larger informed technical community would have if they were to conduct the study"* (NUREG/CR-6372). This statement is so central to the SSHAC process that it is worthwhile briefly discussing each part.

One key objective of a Probabilistic Seismic Hazard Analysis (PSHA) is to capture the full range of possible estimates of the seismic hazard at a site. The hazard is a characterization of nature and, therefore, is theoretically knowable. In reality, however, a level of existing epistemic uncertainty (see Appendix Section B.3.4) means that a legitimate range of estimates exists within which the actual hazard is believed to lie. The assumption of the SSHAC process is that a reasonable range of estimates of the hazard can be captured through development, assessment, and weighting of the scientifically justifiable and defensible interpretations of earth science and geotechnical data by appropriate experts in these fields. The key feature of a SSHAC process is the interaction among these experts as they make and revise their interpretations, which is discussed in greater detail in Sections 3.3 and 3.5.

Once a group of geological, seismological, and geotechnical experts have made their evaluations of all of the available data, the center of these interpretations can be thought of as the best estimate or central value (median) of the distribution of possible outcomes as determined by that group. The term "body" can be thought of as the shape of the distribution of interpretations that lie around this best estimate and capture the major portion of the mass of the distribution. The term "range" refers to the tails of this distribution and the limiting credible values.

The SSHAC process seeks to capture the center, the body, and the range on each component of the hazard study (geographical limits of seismogenic sources, seismic activity rates, maximum magnitudes, ground-motion prediction equations, etc.). If the correlations between these component distributions are also captured, this in turn will then result in capture of the center, the body, and the range of seismic hazard estimates, which is the ultimate objective of the process. In other words, at the specific site of interest, the process will yield the center, the body, and the range of estimates of the annual frequency of exceedance of different levels of each ground-motion parameter. This is the information required for undertaking risk-informed design and evaluations for seismic safety of critical facilities.

The commentary above has focused on the center, the body, and the range of legitimate scientific interpretations of available data as made by a particular group of experts. However, the statement from the original SSHAC guidelines cited at the start of this section refers specifically to the center, the body, and the range (or CBR) of the *informed* technical community. The concept of the informed technical community (ITC) is fundamental to the SSHAC process, but it can be misunderstood. One way that confusion has been generated about the ITC concept is by ignoring the qualifying statement[7] that the process seeks the CBR that the ITC *"would have if they were to conduct the study,"* to which perhaps should be added *"including going through the same interaction process."* The key word in the concept of ITC is "informed," which is specifically defined in the SSHAC guidelines to mean an expert who has full access to the complete database developed for a project and has fully participated in the interactive SSHAC process. In other words, the experts who participate in the particular PSHA study are tasked to represent *"the larger informed technical community"* by assuming the hypothetical case where the others in the larger technical community become "informed" through participation in the same assessment process. There is no pejorative implication that members of the technical community outside the project (i.e., uninformed) are less qualified; it is simply that those who do not participate in the complete SSHAC process are not informed by the databases and interactions specific to the project.

The intent of the SSHAC concept of capturing the CBR of the ITC is admirable; however, it demands that the experts on any given study think about their own interpretations of the available data and also consider the views of the larger technical community. Moreover, it asks the experts to consider what the views of the larger community would be if it had gone through the same SSHAC process. The SSHAC guidelines recognize that this is a hypothetical exercise, but the goal would be to ensure that a broad range of views is considered. In practice, however, often the term "informed" is either ignored or interpreted as simply meaning expert in the field of interest. Thus, the process of capturing or representing the CBR of the ITC has been viewed by some as a process of somehow conducting a poll or surveying the larger community for their opinions. The CBR of legitimate technical interpretations of earth science and geotechnical data is <u>not</u> captured by taking a poll of all suitably qualified and highly regarded researchers and practitioners in these fields. Such an opinion poll is not the objective of a SSHAC process, firstly because a critical element of the SSHAC approach is that the experts make their judgments and evaluations within a structured process that includes full access to all available data. Secondly, the process is not a poll because it is intended to obtain expert judgments (which require a line of reasoning from evidence to claim) as opposed to opinions.

One way to ensure that a PSHA study captured the CBR of the larger ITC would be to conduct the study with the participation of *all* suitably qualified experts in the field, but this would clearly be impractical. In reality, a relatively small number of experts will actually participate in any PSHA study for purely pragmatic reasons, but the process is designed such that the selection of these individuals (provided they have desirable attributes of an evaluator as listed in Section 3.6.3) is not critical to successfully meeting the fundamental objective of capturing the CBR of the ITC. The selected expert participants must possess a sufficiently strong technical background in the relevant discipline (e.g., geology, seismology, engineering seismology, geotechnical engineering) if they are to act as technical evaluators. The experts also must have the personal qualities that make their participation constructive rather than disruptive to the process. This essentially means that experts must be open-minded enough to effectively receive new information, objectively evaluate it, and update their own views in the light of this information as well as be willing and able to participate in scientific debate in a courteous and nonconfrontational manner. Provided the experts meet the criteria in terms of technical background and personal qualities and keep in mind that they must represent the broad views

---

[7] As provided in the quotation in the opening paragraph of this section

of the ITC, then in theory the SSHAC process is expected to produce the same outcome no matter which subgroup of the broader technical community actually participates in the project.

Any selected group of suitable experts would be expected to produce the same outcome in terms of CBR of the final hazard estimates from a SSHAC process. This is because they are charged with considering the views of the larger technical community and, ultimately, with developing models and assessments that capture the CBR of technically defensible interpretations of the available data, models, and methods. The experts are not charged with providing merely their own distribution of technical interpretations. The SSHAC process envisages that the experts will generally do this in two stages called *evaluation* and *integration*, which are discussed further below.

The *evaluation* process starts by the Technical Integrator (TI) team identifying, with input from resource and proponent experts, the available body of hazard-relevant data, models, and methods—including all those previously produced by the technical community—to the extent possible. This body of knowledge is supplemented by new data gathered within the project. The first workshop assists with the identification of hazard-relevant data. The TI team then evaluates these data, models, and methods and documents both the process by which this evaluation was undertaken and the technical bases for all decisions made regarding the quality and usefulness of these data, models, and methods. This evaluation process includes interaction with and (at Levels 3 and 4) among members of the technical community. The interaction includes subjecting their data, models, and methods to technical challenge and defense. Workshop #2 provides a forum for proponents of alternative viewpoints to debate the merits of their models. The successful execution of the evaluation is confirmed by the concurrence of the PPRP that the TI team has provided adequate technical bases for its conclusions about the quality and usefulness of the data, models, and methods, and has adhered to the SSHAC assessment process. The PPRP also will provide guidance regarding meeting the objective of considering all of the views and models existing in the technical community.

Informed by this evaluation process, the TI team then performs an *integration* process that may include incorporating existing models and methods, developing new methods, and building new models. The objective of this integration process is to capture the CBR of technically defensible interpretations of the available data, models, and methods. The technical bases for the weights on different models in the final distribution as well as the exclusion of any models and methods proposed by the technical community need to be justified. Workshop #3 provides an opportunity for the experts to review hazard-related feedback on their preliminary models and to receive comments on their models from the PPRP. To conclude the project satisfactorily, the PPRP also will need to confirm that the SSHAC assessment process was adhered to throughout and that all technical assessments have been sufficiently justified and documented.

Therefore, consistent with the original intent of the SSHAC guidance, we recast the goals of the SSHAC process in terms of the two main activities—evaluation and integration—by the following statement:

> The fundamental goal of a SSHAC process is to properly carry out and completely document the activities of evaluation and integration, defined as:

> *Evaluation:* The consideration of the complete set of data, models, and methods proposed by the larger technical community that are relevant to the hazard analysis.

Integration: Representing the center, body, and range of technically defensible interpretations in light of the evaluation process (i.e., informed by the assessment of existing data, models, and methods).

In light of these definitions, we propose that it is clearer to refer to the CBR of the "technically defensible interpretations" (TDI) instead of the CBR of the ITC. It is important to emphasize, however, that the careful evaluation of the larger technical community's viewpoints remains a vital part of the SSHAC process. We simply have removed the term "informed" because of its specialized definition in the original SSHAC guidelines. Similarly, we propose to replace the term "community distribution" that is used frequently in the original SSHAC guidelines to describe the outcome from a SSHAC assessment process with the term "Integrated Distribution." This will remove any perception that the final assessments and models were arrived at through some type of poll of the community.

Section 5.7 discusses the difficult question of how (and if) one can verify that the CBR of the ITC (or TDI) has been captured.

3.2 Intellectual Ownership of Component Models and Hazard

The original SSHAC guidelines stated that *"it is absolutely necessary that there be a clear definition of ownership of the inputs into the PSHA (and hence ownership of the results of the PSHA)"* (NUREG/CR-6372). Regardless of the SSHAC Level of the study (see Section 4.2), the definition of ownership is indeed extremely important both because it focuses the attention of the project participants on the importance of their contribution to the overall product and because it is key to the process involving multiple experts.

The discussion necessarily begins with what is meant by "ownership" in this context. The project sponsor (i.e., the organization funding the study) owns the results of the study in the sense of property ownership. In other words, the project sponsors have legal ownership of the project deliverables. This is quite distinct from the intellectual ownership of the results, which means taking responsibility for the robustness and defensibility of the various inputs to the hazard calculations. Intellectual ownership of these inputs therefore implies being able and willing to provide full and detailed explanation of the technical bases for all the decisions that led to the models and parameter values entered into the hazard calculations as well as for the rationale behind the relative weighting of these choices on the branches of the logic-tree.

The distribution of intellectual ownership (hereafter referred to simply as ownership) varies with the SSHAC Level at which the study is being conducted. In a Level 1 or Level 2 study, the insistence on clearly defined ownership means that the person or team conducting the study cannot simply adopt and include a model or a parameter value proposed by someone else without justifying its technical basis. In other words, the analyst may incorporate models and parameters proposed by others but then must assume ownership of those models and parameters as part of their own hazard input.

In a Level 3 study, membership in the TI team (see Sections 3.3 and 3.6.4) automatically implies sharing the ownership of the component models developed by that team. By the end of the project, each member of the TI team must be willing and able to speak for the full distribution of component models including their technical content and how the models, taken as a whole, represent the CBR of the TDI. If this is not the case, the process has not been fully successful and additional discussions are needed to arrive at a final distribution for which each member of the team would be willing to provide a technical defense. Clearly, it is helpful if each individual joining the TI team understands from the outset of the project that this is the desired objective.

In a Level 4 study, the evaluators act as individual members of an expert panel with each member of the panel delivering an individual distribution of models and parameters[8] as input to the PSHA. These individual expert assessments must then be combined to arrive at an integrated distribution across the panel. In describing acceptable combining schemes, the original SSHAC guidelines allow the Technical Facilitator Integrator (TFI) to assign unequal weights to the individual evaluator experts' assessments. As discussed in Section 2.2 and stated in the SSHAC report, this decision was motivated by the potential need to deal with "'outlier experts,' experts who make interpretations that are significantly different than those of the rest of the panel and that are not well supported by logic or data" (NUREG/CR-6372, p. 34). While acknowledging this possibility, the SSHAC guidelines encourage the establishment of the proper conditions that would support assigning equal weights including providing access to the same datasets to all participants and ensuring that all experts assume their roles as evaluators. Thus, the notion of unequal weights should only be invoked if an expert clearly refuses to adopt the role and responsibilities of an objective evaluator and if attempts to address this issue during the course of the project are unsuccessful. It should be acknowledged that to some extent the fact that the TFI can assign a low (or, in the extreme case, zero) weight to the assessment of any individual evaluator expert can undermine the sense of ownership on the part of the evaluators.

The existence of the role of the TFI (see Sections 3.3 and 3.6.6) in a Level 4 study also creates a potential vulnerability because it can lead individual experts to believe that their charge is limited to ownership of their own distribution of models only and not to the integrated distribution across the entire expert panel. This is generally not the case and is a pitfall that must be avoided because it effectively diminishes the perception of the importance of these individual expert assessments. The SSHAC process requires a high degree of interaction among the evaluators and between the evaluators and the TFI. The interaction among the parties should include technical challenges to the proposed distributions and their technical bases such that all evaluators understand the proposed distributions. In these exchanges, the evaluators should begin to share a sense of ownership of the integrated model formed by all the individual expert assessments. By the end of the process, the ownership should be fully shared by the expert panel and the TFI. This means that the team in aggregate owns the full model and can explain its technical content. In other words, each member of the panel should agree that the full suite of assessments, taken as a whole, is representative of the CBR of the TDI.

3.3 SSHAC Levels and Relation to Regulatory Assurance

The original SSHAC guidelines defined four levels at which a PSHA can be conducted, with the number of participants, resources, and time required increasing as one progresses from Level 1 to Level 4. The guidelines also envisaged that the majority of hazard analyses would continue to be conducted at the lower levels (1 and 2). The SSHAC guidelines actually make an analogy between the numbers of studies conducted at different levels and magnitude-recurrence relationships for earthquakes. In reality, this analogy probably underestimates the relative proportion of studies carried out as Level 1 and 2 processes.

---

[8] As noted in Section 3.1, models and parameters should be developed to represent an evaluation of the views of the larger technical community and their integration of technically defensible interpretations.

The original SSHAC guidelines conveyed the impression that the biggest step from one level to another is between Levels 3 and 4 because Levels 1, 2, and 3 are all based on the concept of a TI (see Section 3.6.4) whereas a Level 4 study introduces the concept of the TFI. In practice, the important differentiation in terms of complexity, cost, and schedule is between the simpler processes of Levels 1 and 2 and the more involved processes of Levels 3 and 4, which both include workshops, the participation of a participatory peer review panel, and generally larger groups of evaluators. The differences in structure and procedure of a Level 4 study create the need for the role of Facilitator (see Section 3.6.6), which is an important distinguishing feature.

In a Level 1 or Level 2 study, the TI may be an individual analyst or a small team of analysts working together. The TI acts as both an evaluator expert (Section 3.6.3) and an integrator (Section 3.6.4) in developing a logic-tree for the PSHA. The evaluations will be based on published and unpublished datasets, models, and parameters to which the TI is able to gain access. The technical bases for the logic-tree, in terms of branches and weights, must be documented and peer-reviewed by one or more experts external to the project. The SSHAC guidelines strongly recommend that the peer review should not be conducted only at a late-stage (i.e., a review of a draft of the final report) because this generally does not allow any errors to be corrected or significant adjustments to be made. The peer review can more usefully occur in at least two phases—the first focusing on the logic-tree on which the PSHA calculations are to be based and the second on the final documentation of the input and the results.

The feature that distinguishes a Level 2 study from a Level 1 study is that in the former the TI communicates with members of the technical community who have developed the datasets or models that have formed the bases for the logic-tree. The main purpose of this communication is to obtain greater insight into the nature of datasets and how they were compiled, or into the development of models and any assumptions that this involved. Another motivation for this communication is to assess the level of support among the technical community for different alternative hypotheses. Such interactions are likely to be required when considerable uncertainty or controversy exists regarding one or more elements of the seismic source characterization (SSC) or ground motion characterization (GMC) models, particularly if these are likely to exert a pronounced influence on the final hazard results. Such interactions between the TI and data providers or model developers in the technical community will provide the TI with information that is used to identify technically defensible models and parameters and include them in the logic-tree. These interactions also help the TI to assign weights that are based, in part, on the degree of support for each branch given the available data, models, and methods. Most importantly, communication with members of the larger technical community supports the goal that the views of the larger community are considered and increases the likelihood that the hazard assessment is capturing the CBR of the TDI.

In moving up to a Level 3 study, a number of distinct and important changes occur with respect to Level 1 and 2 studies, the main ones being as follows:

i. The TI must now be a team rather than an individual or small group, and a TI Lead (Section 3.6.5) should be designated for purposes of coordination and leadership. Even at Level 1 or 2, the TI might be a small team rather than an individual. However, at Level 3, this is essential both because no individual has the breadth of expertise required and because of the necessity for technical challenge and defense among the evaluators. The team must collectively cover all relevant scientific disciplines, and it is sometimes decided to form separate TI teams for the SSC and GMC components, each with its own corresponding TI Lead.

ii.   A PPRP is formed and provides formal technical and process review throughout the project. The members of the PPRP must collectively possess expertise covering all of the technical disciplines important to the project. The PPRP members also must have practical experience in undertaking seismic hazard assessment. At least some of the members of the group also should have a thorough understanding of the SSHAC guidelines. Section 3.6.8 discusses the PPRP in more detail.

iii.   The project will include a minimum of three formal workshops, as discussed in Sections 4.2 and 4.6. The workshops are centered around certain themes, the proceedings are documented, the PPRP is present as observers, and a large number of resource or proponent experts participate in the proceedings. Within these formal workshops, the TI team interacts with the invited resource experts (see Section 3.6.1) and proponent experts (see Section 3.6.2) who typically are compensated for their preparation and participation. The remuneration ensures that they are able to dedicate the required time to the engagement, and it reflects the seriousness afforded to their participation. The workshops provide an opportunity for formal presentations by these experts and for the TI team members to explore the technical bases for data sets and alternative interpretations with knowledgeable experts. In a Level 2 study, such interactions are informal, and their documentation may be little more than a note to the effect that they took place and a description of the key information that emerged. In a Level 3, the interactions are conducted openly in the presence of observers including the PPRP.

iv.   Between the workshops, the TI team conducts its evaluation and integration processes with the assistance of multiple working meetings. These meetings provide the opportunity for the team to identify and evaluate the data, models, and methods that exist within the larger technical community. These meetings also provide an opportunity to build models that represent the CBR of technically defensible interpretations.

v.   The project documentation includes a full description of the SSHAC Level 3 process followed, all of the information and views presented at the workshops, and a complete and transparent explanation of the decisions of the TI team with regard to the evaluation and integration processes that they have followed. The documentation also must fully describe the basis for the inclusion or exclusion of data, models and methods as well as the weights in the final model. The hazard calculations and sensitivity analyses also are included, along with any project-specific deliverables. The documentation also will include reports from the PPRP written after each workshop and the formal response from the TI Team as well as the evaluation by the PPRP of the final report.

The final logic-tree is developed within the TI team, but at intermediate stages of the project its development is presented and discussed at the workshops.

In a Level 4 study, features (ii), (iii), and (v) listed above are also present, but for item (i) the TI team is replaced by a panel, or several panels, of evaluator experts (see Section 3.6.3) and the TI Lead is replaced by a TFI). The key difference between the TI Lead and TFI roles is the "facilitation" in a Level 4 study whereby the TFI encourages and facilitates the development of individual assessments by the panel members and then structures interactions or discussions among the panel members regarding the technical justification of their individual assessments. Generally, separate expert panels and TFIs will be available for the SSC and GMC components of a Level 4 PSHA. In addition, differences also exist between Level 3 and 4 in item (iv) above. The evaluator experts develop individual distributions that are presented and subjected to technical challenge both in the setting of formal workshops and sometimes in closed working meetings with the TFI. In the workshops, both the TFI and other evaluators challenge each expert's proposed distribution and its technical bases. In addition, a Level 4 study will include individual meetings between the TFI and each expert evaluator during the integration or model-building part of the project.

Chapter 4 explains the differences in the essential steps between Level 3 and 4 studies in more detail, and Section 4.2 describes issues associated with the selection of SSHAC Levels. Also, Section 3.6.6 discusses the TFI role, which is unique to Level 4 studies. Another difference, discussed in Section 3.6.8, is that the role of PPRP at Level 4 focuses more on process review than review of technical assessments. This is because the interactions among the expert panels and their interaction with the TFI are considered to provide a high degree of internal technical review.

In view of how central the concept of *regulatory assurance* is to these discussions, it is worthwhile to clearly define this term. We define regulatory assurance to mean confidence on the part of a regulator (or reviewer), such as the U.S. Nuclear Regulatory Commission (NRC), that the data, models, and methods of the larger technical community have been properly considered and that the CBR of technically defensible interpretations have been appropriately represented. In other words, it is confidence that the basic objectives of a SSHAC process have been met. We make a distinction with the term "reasonable assurance" because it has a specific definition within the NRC's regulatory framework related to compliance with regulations. In contrast, regulatory assurance is a qualitative term that is specific to the SSHAC process and the confidence that is engendered by its proper execution.

Because adopting a Level 3 or a Level 4 process to conduct a PSHA results is a significant increase in terms of cost and duration of the study over that required to conduct a Level 1 or Level 2 PSHA, it is important to highlight the benefits that can be expected to be gained by moving to these higher levels. These benefits are associated with the greater levels of regulatory assurance in Level 3 and 4 studies. For example, the data identification and evaluation process is more explicit and comprehensive because members of the technical community have directly participated in sharing their knowledge of pertinent databases. One benefit is that the study and its technical bases will be more transparent, having been presented in formal workshops and clearly documented. Of course, Level 1 and 2 studies also should be clearly documented, but the documentation is likely to be more extensive in a Level 3 or 4 study. One reason for this is that a Level 1 or 2 PSHA report will explain the choices of models included in the logic-tree whereas, at Level 3 or 4, a greater onus exists to demonstrate that the full range of data, models, and methods has been considered. Consequently, a need exists to also document the models not included in the logic-tree (and the justification for their exclusion). The decisions involved in constructing the logic-tree for the PSHA will have undergone extensive technical challenge and will be subjected to ongoing review by the PPRP. The result of this increased scrutiny during the process is that the input to the hazard calculations and, consequently, the hazard estimates themselves are less likely to be subsequently challenged or shown to be deficient. In addition, the participation of several well-regarded technical experts acknowledged as authorities in their respective fields and interacting within the formal constraints of a Level 3 or 4 process increases the likelihood that the full spectrum of knowledge and uncertainty in the inputs—and therefore the full range of uncertainty associated with the hazard—have been represented. For these reasons, only Level 3 and 4 assessments are appropriate for the basis of PSHA used to develop design levels for nuclear facilities.

A desirable outcome of a Level 3 or 4 study is increased longevity and stability of the hazard assessment. This means that the numerical results of the hazard analysis can be expected to remain stable for a reasonable period of time after the completion of the hazard study. Of course, the appearance of significant new information—such as an earthquake larger than anticipated, the discovery of a previously unknown active fault, or a collection of ground-motion recordings that fundamentally contradict all current models—at any time can lead to the necessity to revisit the hazard analysis. However, such a revisitation is far less likely to be required in a Level 3 or Level 4 study as a result of the significant efforts to identify all existing information and models.

Additional assurance that the range of technically defensible interpretations has been appropriately represented may be provided in a Level 4 assessment because evaluations by individual experts or teams of experts lead to a suite of separate models that in aggregate constitute the final integrated distribution. In terms of how the process is perceived from outside the project, this may be helpful compared to a Level 3 study where the way individual expert assessments contribute to the final composite distribution may be less obvious. The choice of a Level 4 study may lead to greater confidence that the CBR of the TDI has been captured because of the number of individual logic-trees—each of which attempts to capture the full range of uncertainty—combined in a composite logic-tree. This contrasts with a Level 3 study where the individual contributions and models of the members of the TI team are not discernable in the presentations at workshops and in the final report. This arises because the members of the team bring their individual assessments to working meetings and—through discussion, challenge, and defense—these assessments are integrated into a single consensus logic-tree.

A stable assessment of the hazard at a site—determined by experts in a transparent process under the continuous review by a panel of separate but equally experienced experts—provides greater assurance to the NRC (or another regulator) that uncertainties have been effectively captured. In turn, a strong hazard study provides the underpinnings for determination of the design basis ground motions for a critical facility. The technical basis for the PSHA can be more easily defended should contentions be raised. This increased regulatory assurance is the primary benefit obtained by conducting a Level 3 or Level 4 study. However, it is very important to emphasize that adoption of a Level 3 or 4 process does not guarantee regulatory acceptance even if the project fully conforms to the procedural requirements.

3.4 Data Collection Versus Expert Judgment

The nature of hazard analyses for rare, large-consequence events—coupled with the wide range of seismological, geological, strong-motion and geotechnical data that can be used for these analyses—is such that scientists will generally be able to make several legitimate and defensible interpretations from the set of available data. These scientifically viable alternate interpretations represent the epistemic uncertainty that must be captured in the logic-tree. The reason for conducting multiple-expert assessments is to increase the likelihood of capturing the full range of this uncertainty. A vitally important issue here, sometimes overlooked in planning large PSHA studies, is that such expert judgment should only be used to identify and quantify the uncertainty that remains after appropriate data collection, compilation, and evaluation activities have been completed.

A distinction is made here between data *compilation* and data *collection*. Data compilation involves the assembly of all pertinent data that exist at the time a hazard analysis is conducted. Compiled data are entered into the project database and form the basis for the evaluation and integration processes. Data collection involves conducting new scientific studies beyond those that are available in the technical community. In most cases, new data collection activities for a SSHAC project are highly focused on issues of importance to the hazard analysis and are specifically designed to reduce uncertainties in the key inputs to the hazard analysis. Moreover, these studies are completed in a timeframe that allows for their use in the hazard study.

Although new data is not required to be collected within the SSHAC guidelines, it should be considered because of the potential for new data to reduce key uncertainties in the hazard inputs. In some cases, however, the NRC or other regulatory bodies require specific data collection activities. An advantage of a well-structured SSHAC process is that it can be used to identify specific data collection activities that have highest potential to reduce the most hazard-significant uncertainties. One of the first steps in a SSHAC Level 3 or 4 process is the identification of hazard-significant issues and the data that are available to address those issues (Section 4.5). If significant data gaps exist, new focused data collection efforts can be considered. Geological data collection and processing can include, for example, field studies, interpretation of remote sensing imagery, geodetic measurements of deformation rates, gravity and magnetic surveys, trenching, and dating of deposits. Seismological investigations can include the collection of historical information related to pre-instrumental earthquakes, the collection and reprocessing of seismograms from the early instrumental era, relocation of earthquake hypocenters, the use of improved velocity models and advanced algorithms, and (of course) the installation of additional instruments. Strong-motion data collection and processing activities also can include instrumentation of the site or region (even weak-motion recordings from the site of interest can be of great value) and the geotechnical characterization of recording stations as well as testing of predictive equations using local strong-motion datasets and intensity observations.

Such new data collection activities are important because the expansion of datasets can reduce uncertainty in an assessment (although it is important to bear in mind that it might not be smaller than the apparent uncertainty in earlier assessments as discussed in Section B.4). The smaller the uncertainty, the more useful and more robust will be this characterization. Because uncertainty will always be present to some degree, it is very important that it is captured in the assessment. The SSHAC process is designed to assist the analyst in meeting the objective of capturing the state of knowledge and associated uncertainties regarding the seismic hazard at a given location at a given point in time. Therefore, multiple-expert assessment should never be used as a substitute for data collection. In other words, the experts in a PSHA should never be used to infer or guess values that could reasonably be measured within the time and budget resource constraints of a project. To do so is a misuse of the SSHAC process.

Acknowledging the benefits of new data collection, a trade-off exists between the resources required to conduct new data collection activities and the potential for uncertainty reduction. Situations will occur where pressures of budget and schedule preclude acquisition or collection of new data; consequently, attention should be paid to the minimum data collection requirements specified in regulatory guidance such as RG 1.208 (USNRC, 2007). However, even in cases where budget and schedule constraints preclude collection of new data, comprehensive data compilation (i.e., gathering and ordering all existing information) is an indispensable core requirement. The project documentation must provide evidence that all relevant data were compiled as part of the project. The documentation also must demonstrate that the evaluator experts carefully considered the resulting dataset in their assessments. In a Level 3 or Level 4 study, one of the key mandates of the PPRP is to ensure that this is done.

3.5 Multiple-Expert Interaction Versus Expert Elicitation

The original SSHAC guidelines make repeated reference to "expert elicitation" in describing the SSHAC process. As discussed in Section 2, the use of this terminology derives from the history of the use of multiple experts in seismic hazard and risk analyses whereby decision analysis techniques are used to elicit the judgments of experts. However, it is more appropriate to refer to the SSHAC process as multiple-expert assessment because it differs from classical expert elicitation in a number of ways. One key difference is the nature of the output required from the experts and the definition of "expert" that this implies. Another important difference is the fact that, in SSHAC-based studies, experts are expected to interact rather than remain independent from one another. The other respect in which the SSHAC process is distinct from classical expert elicitation is that the assessments of individual experts are integrated rather than aggregated. The following paragraphs discuss each of these differences in more detail.

In classical expert elicitation, the objective is to obtain answers to well-defined questions from carefully selected experts. In such a process, the answers are assumed to already exist in the minds of the experts and so the goal is skillful extraction of the expert assessments combined with calibration of the experts to ascertain which answers constitute the most reliable data. In other words, the experts are perceived as repositories of knowledge, and it is assumed that the knowledge only needs to be elicited from them. This sits in marked contrast to the SSHAC process in which evaluators, chosen in part because of their appropriate technical backgrounds, actually become "experts" in the course of the project as they evaluate project-specific data and learn from the interactive process. Subject matter experts are asked to participate in an interactive process of ongoing data evaluation, learning, model building, and, ultimately, quantification of uncertainty. Through the process, they are required to develop judgments or interpretations for which they must provide technical justification. The experts must defend their evaluations in the face of technical challenge from other participants, and they are also expected to challenge and interrogate the other evaluators. All of this is intended both to assist the experts as they endeavor to give full consideration to the data, models, and methods of the broader technical community of which they are part, and to ensure that those views are appropriately represented in the resulting analyses.

Classical expert elicitation is often focused primarily, if not exclusively, on eliciting probabilities for events or outcomes presented to the experts by those conducting the elicitation (e.g., O'Hagan et al., 2006). Therefore, while they may be selected on the basis of their expertise in a particular field (substantive expertise), the role of the experts in the process becomes more closely related in many ways to their normative expertise (ability to provide coherent and unbiased probability assessments). Indeed, in many ways the relative success of an expert in a classical elicitation may be heavily influenced by their ability to objectively judge their own biases. In a SSHAC process both attributes (substantive and normative expertise) are important. However, a large part of the task of an evaluator expert is related to technical assessments in terms of actually selecting, and in many cases adapting or even developing, models for the specific application at hand (see Section 5.6). Assigning weights (which are subjective probabilities in terms of how they are subsequently treated) to the selected and adjusted models is not the entire process. In relation to this, Section B.3.5 discusses insights related to the relative insensitivity of the hazard results to some branch weights in a logic-tree.

Whereas in classical expert elicitation emphasis is placed on maintaining the strict independence of each expert, to avoid "cross contamination," the SSHAC process expressly encourages and fosters structured interactions among experts during the assessment process up to and including discussion of preliminary assessments of specific uncertain quantities. The purpose is not to achieve consensus although, if this occurs through genuine convergence of the expert assessments, it is an acceptable outcome. The interactions among evaluator experts—apart from providing additional technical challenge—serve two important purposes. First, they deepen understanding of the problem and data in the course of the joint learning experience. Second, they ensure that at the end of the project any remaining differences among the assessments of individual experts represent genuine epistemic uncertainty and do not result from misunderstandings or from exposure to different sets of data or models. This is especially important in seismic hazard analysis because the total data available for the assessments is often limited because of dealing with rare events. Thus, there is simply no way that experts can be truly independent and rely on their own unique data sets.

The final major difference between the SSHAC process and classical expert elicitation relates to the way in which the judgments of individuals or teams of evaluator experts are subsequently combined at the end of the process. In standard expert elicitations, these individual judgments are aggregated using various combination rules that relate to the "quality" or degree to which each expert is "calibrated" relative to known quantities. In a typical expert elicitation, subject matter experts are asked narrowly defined questions about specific uncertain quantities within their area of expertise, and they provide their judgments in the form of probability estimates or distributions. For example, a climate scientist might be asked to provide an estimate of "the equilibrium change in global average surface temperature" given a specific set of circumstances (Morgan and Keith, 1995). In this approach, experts are treated as independent point estimators of an uncertain quantity, and the elicitation "problem" is viewed primarily in terms of determining how to ask the right questions as clearly as possible of the most knowledgeable experts. Based on this perspective, the elicitors may focus significant effort on ensuring that they have well-calibrated and informative experts (i.e., experts who are able to give both accurate estimates and a narrow range of uncertainty in estimates for quantities similar to those of interest in the elicitation but for which a "true" value can be determined) (e.g., Chapter 10 of Bedford and Cooke, 2001). For narrow assessment tasks, the elicitor may focus on designing elicitation questions to motivate "honest" responses through the use of proper scoring rules (Gneiting and Raferty, 2007). All of these tools and approaches reflect the general philosophy that probabilities are something that exist in the experts' minds, and the job of the elicitor is to extract, or elicit, those probabilities. Typically, aggregation of expert assessments, whether through mechanical or behavioral approaches, is conducted for independent estimates of a quantity of interest. Issues inevitably arise regarding such things as the number of samples needed to faithfully represent a "complete" representation of the uncertainties in that quantity (see discussion in Section 2.1).

The SSHAC integration process takes a different view. Evaluator experts examine the available data, probe the technical bases for proponent experts' models, and consider the views of the larger technical community. In the integration step, they build models and quantify uncertainties that represent not just their own views but the CBR of technically defensible interpretations. In this way, all experts in a SSHAC Level 4 or on a Level 3 TI team study are required to assess the same thing—the CBR of the TDI. As a result, the expert elicitation issues related to the proper numbers of experts on a panel and proper "sampling" of the larger technical community are not applicable to a SSHAC process.

3.6 Key Roles and Their Responsibilities and Attributes

SSHAC Level 3 and 4 processes provide a structured and transparent framework for conducting multiple-expert assessments that effectively capture epistemic uncertainty in hazard analyses. Central to the success of the process is the clear definition of the different roles that experts play and how they interact in the process. In the following subsections, each of the different expert roles is described in terms of the role of the individual or group in the process, the responsibilities that the individual or group assumes in the project, and the attributes that an individual should possess to contribute effectively in that particular position. In the following subsections, attributes are often referred to as being required or necessary. In some cases, they may be considered desirable rather than indispensable, but a project is more likely to function successfully the more each participant is able and willing to conduct themselves in accordance with the attributes of their assigned role.

3.6.1 Resource Expert

The *role* of a resource expert is to present data, models and methods in an impartial manner. The resource expert will make this presentation in the setting of a formal workshop in a SSHAC Level 3 or Level 4 study. The expert is expected to present their understanding of a particular data set, including how the data were obtained, or to present a model or a method with their limitations and caveats. In all cases, a resource expert is expected to make the presentation without any interpretation in terms of hazard input. The reason for this is that they are not playing the role of proponents or advocates of particular models or methods.

The main responsibility of a resource expert is to share their technical knowledge in an impartial way in their presentations to the evaluator experts. This means that their presentation should make full disclosure including all caveats, assumptions, and limitations. The resource expert is also expected to respond candidly and impartially to questions posed by the evaluator experts. A resource expert has full responsibility for the material that they present but does not participate in any way in the ownership of the hazard models.

The necessary attributes of a resource expert are knowledge and impartiality. Resource experts must possess a deep and broad knowledge of the tectonics, geology, or seismicity of a particular region (or a data set, model, or method) and will often have worked on that topic for many years and have a number of publications related to the subject of their presentation. They must be able to withhold their judgment with regard to hazard implications of the material that they present.

3.6.2   Proponent Expert

The role of a proponent expert is to advocate a specific model, method, or parameter for use in the hazard analysis. The expert will advocate the model within the forum of a formal workshop. The proponent may be invited to present a model, which will usually be their own, either because the model has been published, is widely known, and is therefore considered a credible option or because the model is controversial. In some cases, a proponent may be invited to present a relatively new model, which may not have even been published at that stage, if it is thought that the model is directly relevant to the hazard analysis and credible.

The responsibility of a proponent is to promote the adoption of his or her model as input to the hazard calculations. The proponent is required to justify this assertion, to demonstrate the technical basis for the model, and to defend the model in the face of technical challenge. The proponent also is charged with making full disclosure about the model in this process including all underlying assumptions. A proponent expert has full responsibility for the material that they present but does not participate in any way in the weighting of alternative hypotheses or in the ownership of the hazard models.

An individual who has another role in the project (such as resource expert or even a member of the evaluation team) could adopt the role of a proponent expert at a specific moment during the project. This would require that everyone present is made very clearly aware of the switch of roles and that (in the case of an evaluator) the individual is prepared to subsequently revert to the role of impartial evaluator. This flexibility is useful not least because in practice individuals have been willing to reveal in this way specific weakness or limitations of models that they developed and, in some cases, even to propose that their models should not be adopted.

The attributes required of a proponent expert are knowledge (the same criteria as described for a resource expert) and the ability to defend his or her model and its basis.

3.6.3 Evaluator Expert

The evaluator expert plays the most important role in a SSHAC process. The role of the evaluator expert is to objectively examine available data and diverse models, challenge their technical bases and underlying assumptions, and—where possible—test the models against observations. The process of evaluation includes identifying the issues and the applicable data, interacting among the experts (i.e., challenging other evaluators and proponent experts, interrogating resource experts), and finally considering and weighing alternative models and proponent viewpoints. In Level 3 and 4 studies, evaluators and integrators (see Section 3.6.4) are the same people, but these labels refer to distinct roles played at different stages of the project.

The responsibility of the evaluator is to identify existing data, models, and methods as well as alternative technical interpretations and to evaluate these in terms of their general quality/reliability and their specific applicability to the assessments being made. In the second phase (which is integration discussed in Section 3.1), the evaluator is charged with constructing a logic-tree using existing and/or new models and assigning branch weights so that they reflect his or her understanding of the broader community distribution. The evaluator must present a clear defense and rationale for their choices both in terms of selected models and the weights assigned to them. The evaluator is not obliged to include in the logic-tree all proponent viewpoints but must provide documented justification for excluding any particular model. In a Level 3 project, the evaluator executes this responsibility as part of a team and, therefore, must interact openly and constructively with the other members of the team.

The attributes required for an evaluator expert include possession of a strong technical background, the ability to objectively evaluate the strengths and weaknesses of alternative models, and at least some familiarity with approaches to quantifying uncertainties for hazard analysis. The strong technical background is required to enable the evaluator to make informed evaluations of existing models and the impact of the models on seismic hazard. For this reason, evaluator experts also should have an understanding of the basic mechanics of PSHA and how the elements that they are charged with evaluating influence and impact upon the hazard estimates. The last criterion, however, should not preclude the selection of an otherwise suitable expert because the project leaders can provide instruction on these issues as part of the process.

In a Level 3 study, evaluator experts also must have the ability to work in teams, and a congenial and respectful approach to other team members is obviously very desirable. At the same time, an evaluator expert should have good communication skills to challenge proponent views and to defend their own assessments. In addition, an evaluator must be able to act with objectivity and be willing to forsake the role of proponent, up to and including critical assessment of models that they may have developed. An evaluator expert also must be able to commit significant time and effort to the project. Whereas resource and proponent experts generally only need to attend one or perhaps two workshops, an evaluator expert must be present at all workshops and relevant working meetings and commit to the entire duration of the project.

Because interaction among evaluators is an integral and valuable part of the process, it is important that evaluators bring different perspectives to the project. For this reason, it is desirable that no single organization or group be heavily represented within the evaluator teams.

3.6.4 Technical Integrator

Integration is the process by which evaluations of candidate models are brought together in a single logic-tree that reflects the CBR of the TDI. The role of TI, as noted above, is linked with the role of evaluator. Explicity, in Level 3 and 4 studies, an evaluator expert adopts the role of integrator during the second phase of their assessment when they develop a model that reflects their understanding of range of technically defensible interpretations. In a Level 3 study, the interactions with other evaluators in the project also contribute to a successful integration process. In a Level 4 study, evaluators contribute to the process of integration in the second phase by assessing the integrated distribution individually. In addition, the individual assessments are combined into a single integrated distribution across all experts on the panel using a particular combination scheme (see Sections 2.1 and 4.10.1).

The responsibility of a TI is to construct a model for input to PSHA that captures the CBR of the TDI and to provide complete and clear justifications of the technical bases for all elements of the model including the reasons for excluding or down-weighting any data, models, or methods. TIs in Level 3 and 4 studies will always be members of a team (Level 3) or a panel (Level 4), and they have the responsibility to challenge the technical basis of assessments made by the other evaluators and to subject their own assessments to the same challenge.

The main attributes of a TI are the ability to objectively evaluate the views of others in developing models and expressions of uncertainty and to deeply appreciate the influences of different models and parameters on hazard results. The TI also needs to be able to produce large volumes of clear and complete documentation on schedule as well as critically review all contributed documentation from evaluators and others. The TI must be willing and able to make a major commitment of time and effort to the project.

3.6.5 Technical Integrator Lead

In a Level 3 project, it will routinely be necessary to appoint a TI Lead to coordinate the activities of the TI team. The main roles of the TI Lead are to coordinate the activities of the team of expert evaluators and to serve as the point of contact with the project manager, other TI Teams (in a full PSHA, with SSC and GMC sub-projects), and the Project Technical Integrator (PTI) (see Section 3.6.7).

The responsibilities of the TI Lead include selection of a team of appropriate evaluator experts. Another responsibility is identification of suitable resource and proponent experts and their invitation to the relevant workshops, including clear instructions of the scope for their participation. In the case of an invited resource or proponent expert being unable to attend a workshop, the TI Lead must ensure that their views are fully represented to the evaluators.

The responsibilities of the TI Lead also include running the workshop sessions and ensuring that all participants clearly understand the workshop objectives, their individual roles, the required output from the workshops, and the implications of the issues under discussion for the seismic hazard analysis. The TI Lead also is responsible for ensuring that all evaluators have full access to all of the available data and information. A key responsibility of the TI Lead is to ensure that the project documentation is complete and comprehensive (see Section 4.10). The TI Lead also has responsibility for ensuring that all members of the TI team are made aware of the potential for cognitive bias and are alerted to when biases may be influencing their assessments. The TI Lead also will be responsible for instructing any members of the TI team who are not fully conversant with the concepts of aleatory variability and epistemic uncertainty and their application to PSHA.

The main attributes of a TI Lead are a very strong technical background in SSC or GMC issues (as appropriate) and experience in the conduct of PSHA studies. It is desirable that the TI Lead has a good standing in the technical community because it is necessary for members of the TI team to view the TI Lead at least as a peer. More than any role described so far, the TI Lead must be willing and able to make a very major commitment of time and effort to the project.

3.6.6 Technical Facilitator Integrator

The position of facilitator was formally conceived as unique to Level 4 studies where it is embedded in the concept of the Technical Facilitator Integrator (TFI). The role of the TFI includes all the elements described in the previous section for the TI Lead plus those specific to the function of facilitation of the evaluator experts (although the TI Lead position also includes a facilitation role within the TI team). The primary role is to facilitate workshop interactions among the experts on the evaluator panel to ensure that all assessments are challenged and adequately defended and that the evaluators act at all times as objective and impartial assessors. The facilitator also ensures that the evaluators consider the views of the larger technical community and ultimately produce models that reflect their assessments of the CBR of the TDI. All of these tasks may be reinforced through closed working meetings between the TFI and individual experts.

The responsibilities of the TFI include all of those outlined above for the TI Lead, with particular emphasis on the selection of the panel of evaluator experts. The facilitator must encourage the evaluators to challenge one another within the workshops in the same way that the TI Lead must perform the same role in working meetings in a Level 3 study.

In a Level 4 study, the role of TFI is of pivotal importance and it becomes imperative to appoint an appropriately qualified individual. The necessary attributes include all of those listed in Section 3.6.5 for the TI Lead. The TFI particularly requires a good understanding of the various cognitive biases that evaluators can be subjected to in their assessments. The facilitator must be someone who is able to pay attention to detail and remain focused during workshop sessions. Moreover, the facilitator needs to have the ability to communicate effectively and clearly and the willingness to challenge and confront participants to fulfill their roles while maintaining a structured and efficient process.

In view of the demanding list of required attributes, it will generally be the case that a small team will execute the TFI role rather than an individual in a Level 4 project. If a TFI team is established, there is value in identifying a TFI Lead (or lead TFI) to establish a primary point of contact and to have clear lines of communication.

The TFI role is probably the single most demanding position in a SSHAC study. For this reason, it is essential that anyone assuming this role be willing to commit a great deal of time to the project and be more or less continuously available throughout the project duration. In common with the TI Lead, this means, among other things, that the TFI can be relied upon to provide clear and complete responses to questions and requests from the evaluator experts in a timely manner.

3.6.7 Project Technical Integrator and Project Technical Facilitator Integrator

If the project is a full PSHA including both SSC and GMC subprojects or another type of hazard study involving multiple major components, then establishing the position of a Project TI Lead or Project TFI is advantageous. These roles were not envisaged in the original SSHAC guidelines, and their inclusion has been suggested by recent practice resulting from past lessons learned in the application of the guidelines. Section 5.2 discusses the roles of PTI and Project Technical Facilitator Integrator (PTFI).

3.6.8 Participatory Peer Review Panel

The Participatory Peer Review Panel (PPRP) is a key and indispensable element of a SSHAC Level 3 or 4 study. Following issue of the final PSHA report, the PPRP must concur that the project has conformed to the requirements of the specified study level and that all technical assessments have been adequately defended and documented. Because the PPRP will be composed of experienced specialists in the field of seismic hazard assessment, this approval should carry considerable weight and can be expected to increase regulatory assurance.

The PPRP fulfills two parallel roles, the first being technical review. This means that the PPRP is charged with ensuring that the full range of data, models, and methods have been duly considered in the assessment and also that all technical decisions are adequately justified and documented. The second role of the PPRP is process review, which means ensuring that the project conforms to the requirements of the selected SSHAC process level. Collectively, these two roles imply oversight to assure that the integration is performed appropriately.

One point that is important to emphasize is that membership of the PPRP is always on an individual basis and not as an affiliate of any organization. Each member of the PPRP in the employ of an organization must ensure that it is clearly understood that they are not representing their employer or organization on the panel but are serving in their own right as a recognized leader in their respective field.

The responsibility of the PPRP is to provide clear and timely feedback to the TI/TFI and project manager to ensure that any technical or process deficiencies are identified at the earliest possible stage so that they can be corrected. More commonly, the PPRP provides its perspectives and advice regarding the manner in which ongoing activities can be improved or carried out more effectively. In terms of technical review, a key responsibility of the PPRP is to highlight any data, models or proponents that have not been considered. Beyond completeness, it is not within the remit of the PPRP to judge the weighting of the logic-trees in detail but rather to judge the justification provided for the models included or excluded, and for the weights applied to the logic-tree branches.

The PPRP has the clear responsibility to be present at all the formal workshops as observers and to subsequently submit a consensus report containing comments, questions, and suggestions. A separate issue is whether the PPRP has the responsibility to be present at all working meetings as well. Some confusion exists regarding this specific point because of the use of the word "participatory" that was introduced to distinguish this type of review from late-stage review. It has since been suggested that a more appropriate adjective might be "continual" (Hanks et al., 2009) to make it clear that the PPRP does not participate in the technical assessments. The SSHAC guidelines also included statements, however, that may have compounded this confusion, including the following: "*The peer reviewers should meet frequently with the TI to review all aspects of the analysis. Their role is to inform the TI of available data and interpretations being made that might have an impact on the...analysis, to express their own interpretations as experts, to examine and suggest refinements to methods and procedures being followed by the TI, and to ensure that a wide range of technical interpretations is being represented*" (emphasis added, NUREG/CR-6372). The two phrases underlined are questionable in that they could be interpreted as implying participation in the expert evaluation process rather than independent review. In addition, arguments have been made that the TI team should be able to operate freely at informal working meetings without feeling that they are under scrutiny; in other words, they should have the freedom to "think aloud" and to thrash out differences of opinion. Equally, both of the underlined statements contradict an important warning issued elsewhere in the SSHAC guidelines—namely, "*that peer reviewers might lose their objectivity as they interact with the project over time*" (NUREG/CR-6372). Because the PPRP can only fulfill its vital role by remaining separate from the evaluation and integration process, it is recommended that the *basic* responsibilities of the PPRP be limited to the following:

- Review of the project plan.
- Review of workshop agenda and lists of invited resource and proponent experts.
- Attendance at all workshops and timely submission of written reports.
- Participation in daily debriefings with projector leaders at workshops.
- Highlight interface issues (Section 5.9) if these are not being adequately addressed.
- Direct challenge of evaluators' assessment at Workshop #3.
- Review of the preliminary SSC and GMC models for capture of the CBR of the TDI.
- Review of the draft final project report.
- Issue of consensus letter report following completion of final project report.

However, provided that the boundaries are maintained and the clear separation of reviewers and evaluators is respected, then one or more representatives of the PPRP may attend the working meetings of the evaluator experts as observers. This could bring benefits of providing information to the entire PPRP via the observer representative at an early stage regarding the manner in which the evaluation and integration processes are being conducted. Such information can assist the PPRP in their later reviews of the bases for the technical assessments and their review of the project conduct and documentation. This is particularly true in Level 3 studies in which technical challenges to various interpretations by evaluators occur in the working meetings, as well as in the workshops.

One more responsibility of a PPRP member is to preserve their independent status throughout the project. On the one hand, this means not being drawn into the technical assessments to maintain objectivity. On the other hand, it involves resisting any temptation to represent the corporate views of the organization to which they are affiliated, because PPRP members must always serve in an individual rather representative capacity on the panel.

The *attributes* of the PPRP can be defined both for individuals and in collective terms for all of the members of the panel as a group. A key requirement is that each member of the group has an understanding of and commitment to the principles of the SSHAC process. In addition, the members of the panel must collectively cover all technical aspects of building SSC and GMC models and of conducting a PSHA. The requirements for technical expertise are particularly emphasized for a Level 3 study. Similarly, it is desirable that the members of the PPRP are highly regarded within the technical communities, again particularly for a Level 3 study where the role of the PPRP is of paramount importance. The members of the PPRP also should be prepared to commit sufficient time to the project to become fully familiar with the issues, data, and models and to be able to review thoroughly the documentation developed.

It is highly recommended that an individual be named as the chair of the PRPP. The role of the PPRP chair is to liaise with the Project Manager and coordinate the panel itself, particularly in relation to the drafting of written reports and organizing pre- and post-workshop meetings of the panel. The *responsibilities* of the PPRP chair include ensuring that the panel is able to arrive at a consensus view in each case, ensure that concerns are communicated clearly and in a timely fashion to the project, and to energetically follow up on these issues if a satisfactory response is not received. Another responsibility of the PPRP chair is to ensure that the panel remains objective by maintaining a suitable distance from the inner workings of the evaluation teams. The attributes of the PPRP chair include a working knowledge of PSHA, experience in SSHAC Level 3 or 4 projects, and being held in high regard as a technical expert in their own right. The ability to maintain congenial relationships within a group while achieving consensus conclusions is another important characteristic required of the chair.

3.7 Structure and Sequence of Process

This chapter has outlined many of the key concepts of a SSHAC process and defined all the expert roles encountered in Level 3 and Level 4 studies. However, to constitute a *bona fide* SSHAC process, it is not sufficient only to include all of these elements. They also must be embedded within a structured sequence of steps. Chapter 4 describes the required steps and their sequence, but it is worth stating here that the focus should always be on meeting the objectives of each step rather than simply executing the steps themselves. In other words the project should be conducted to meet the objectives of evaluation and integration in a transparent, well-documented, and technically sound process rather than simply to meet minimum requirements to have the appearance of a high-level SSHAC process. The combination of the expert roles defined in this chapter and the process described in Chapter 4 is what can contribute to regulatory assurance.

4. ESSENTIAL STEPS IN SSHAC LEVEL 3 AND 4 PROCESSES

Drawing on the experience gained from the history of multi-expert assessments (Chapter 2) and mindful of the key concepts that underlie the Senior Seismic Hazard Analysis Committee (SSHAC) process (Chapter 3), this chapter defines the essential steps recommended to be followed in SSHAC Level 3 and 4 processes. Note that in the following discussion the term "SSHAC Level" is used synonymously with the "Study Levels" given in the original SSHAC report (NUREG/CR-6372, p.25). As discussed in Section 2.4, the history of the implementation of the SSHAC guidance has led to a number of innovations and improvements in the ways that the SSHAC concepts are implemented in actual projects. Often, in fact, the implementation approaches have been customized and tailored to address the specific issues of importance to the particular study. This continual improvement and evolution of implementation approaches (as well as customization of the approaches for project-specific applications) is commendable and should be encouraged. At the same time, a need exists to describe the minimum requirements for a particular study to be a Level 3 or 4 SSHAC project. Provided that these requirements are met, embellishments and project-specific enhancements can be employed.

The following discussion is focused on the elements of Level 3 and 4 studies because these are most appropriate for nuclear and other safety-critical facilities. The elements of SSHAC Level 1 and 2 studies are only briefly discussed, but they are given some attention in the original SSHAC report (NUREG/CR-6372, Section 3.2.2). In general, a technical integrator (TI) conducts these studies (which may include a small team) without workshops or other structured expert interactions. A Level 1 study is typically conducted based on consideration of available data and information that is generally available within the peer-reviewed literature. Level 2 studies include additional communications with members of the technical community to enable the TI to better understand the knowledge and uncertainties that currently exist. A project report (subject to peer review) documents the technical assessments. Because most of the process is not conducted in an environment that is amenable to ongoing peer review like a Level 3 or 4 study, the documentation is the sole basis for the peer reviewers to understand the assessments made, and it must be complete and comprehensive.

4.1 Summary of Essentials Steps

Table 4-1 summarizes the recommended essential steps for a hazard study to be designated as a SSHAC Level 3 or 4 study. Note that following the selection of the SSHAC Level, the subsequent discussions in this chapter will be related to either Level 3 or 4 studies, with appropriate distinctions made between the two levels, as needed.

4.2 Selection of SSHAC Level

The first decision that must be made for a hazard study is the SSHAC Level at which the project will be conducted. The SSHAC guidance calls for the assignment of SSHAC Levels to be made at the level of "issues," which can entail the entire hazard study or individual technical issues (e.g., the recurrence rate for a particular seismic source). The thought was that, perhaps, resources could be conserved by addressing more uncertain or controversial issues using a higher SSHAC Level, and the remaining issues using a lower Level. Experience thus far has shown that it is very difficult to separate individual technical issues within the seismic source characterization (SSC) and ground motion characterization (GMC) components sufficiently to treat them with different SSHAC Levels. In fact, although it is conceivable that the SSC and GMC components could be treated with different SSHAC Levels, no project has done so thus far.

Table 4-1. Summary of Essential Steps in SSHAC Level 3 and 4 Studies

Essential Step	Discussion
1. Select SSHAC Level	• Document decision criteria and process
2. Develop Project Plan	• Includes project organization and all technical and process activities
3. Select project participants	• Includes all management, technical, and peer review participants
4. Develop project database	• Includes compilation of existing, available data • Can include focused new data collection • Data dissemination to all evaluator experts (Level 4) or TI Team members (Level 3)
5. Hold workshops (minimum of three)	Workshop topics: • Hazard-significant issues and available data • Alternative interpretations • Feedback
6. Develop preliminary model(s) and Hazard Input Document (HID)	• Preliminary models developed prior to Feedback workshop • HID provides input to hazard calculations
7. Perform preliminary hazard calculations and sensitivity analyses	• Intermediate calculations should display the impact of elements of the expert models • Hazard calculations should show the significance of all elements of the models • Sensitivity analyses should include the contributions to uncertainties
8. Finalize models in light of feedback	• Feedback provides a basis for prioritizing and focusing the finalization process • Implement expert combination process across all evaluator experts in SSHAC Level 4
9. Perform final hazard calculations and sensitivity analyses	• Should be conducted to develop the required deliverables for subsequent use of the hazard results
10. Develop draft and final project report	• Fundamental documentation of SSHAC process, technical bases, and results
11. Participatory peer review of entire process	• Periodic written reviews of key products and activities • Review of draft report • Final written review of technical evaluations and process used

The factors that enter into the decision regarding SSHAC Level are usually qualitative and subjective to a large extent. For example, the SSHAC guidance does not prescribe a formula for making the decision but identifies a number of considerations that would inform the decision.

The SSHAC guidance calls for evaluating issues relative to several factors that would establish their "degree" or their need for consideration at a higher SSHAC Level (NUREG/CR-6372, p. 24), including:

- The significance of the issue to the final results of the PSHA.
- The issue's technical complexity and level of uncertainty.
- The amount of technical contention about the issue in the technical community.

The SSHAC also identifies "decision factors" that include regulatory concern, resources available, and public perception.

All of these factors play into a classic decision problem of balancing the costs associated with conducting a particular SSHAC Level with the benefits of doing so. In such a decision process, the potential costs and benefits are first identified, and then the sponsors assess the "value" of each cost and benefit (qualitatively or quantitatively) for the particular application being considered. Table 4-2 provides a summary of important attributes of projects conducted at the various Study Levels, based on experience. As discussed in Section 3.3, the most significant differences in most of the study attributes occur between SSHAC Levels 2 and 3, with much less difference between 1 and 2, or between 3 and 4. In addition, the largest differences in regulatory assurance lie between Level 1 or 2 studies versus Level 3 or 4 studies.

The attributes summarized in Table 4-2 are generic and not specific to any particular project location, facility type, or regulatory environment. Therefore, the second step in the decision process involves a consideration of the project-specific factors.

The following project-specific factors are important for selecting the appropriate SSHAC Level:

- Safety significance of facility (e.g., nuclear power plant, high-consequence[9] dam, bridge, conventional building).

- Technical complexity and uncertainties in hazard inputs.

- Regulatory oversight and requirements (e.g., quality assurance requirements, regulations and regulatory guidance in place, monitoring and audit).

- Amount of contention within technical community.

- Degree of public concern and oversight.

- Resource limitations (e.g., time and money).

These project-specific factors provide a basis for evaluating the relative value of the attributes identified in Table 4-2. For example, consider a study for a conventional building with limited public concern and oversight and severe resource limitations. In this case, highest value would likely be assigned to the attributes of lower cost, shorter durations, and minimal management challenges. Lesser value would be accrued to transparency, regulatory assurance, and broader ownership of the hazard models and results. At the other end of the spectrum, consider a new nuclear power plant that is to be sited in an area of complex tectonics under heavy regulatory and public scrutiny. In this case it is likely that the sponsors of the hazard analysis would place high value on attributes such as participatory peer review, broad ownership, enhanced transparency, and higher levels of regulatory assurance.

---

[9] "High-consequence" dams are sometimes also called "high-hazard" dams.

55

Table 4-2. Attributes of Various SSHAC Levels

SSHAC Level	Level 1	Level 2	Level 3	Level 4
Number of participants	• Project Manager • Small TI team • Peer reviewers • Hazard calculation team	• Project Manager • Small TI team • Peer reviewers • Hazard calculation team • Resource experts • Proponent experts	• Project Manager • Project TI • Larger TI team • Peer reviewers • Resource experts • Proponent experts • Data team • Hazard calculation team	• Project Manager • Project TFI • Small TFI team • Panel(s) of evaluator experts • Peer reviewers • Resource experts • Proponent experts • Data team • Hazard calculation team
Interaction	• Limited or no contact with proponent and resource experts	• Proponent and resource experts contacted individually	• Proponent and resource experts interact with TI Team in facilitated workshops	• Proponent and resource experts interact with evaluator experts in facilitated workshops
Peer review	• Late stage	• Late stage	• Participatory	• Participatory
Ownership	• TI Team	• TI Team	• TI Team	• TFI team and evaluator experts
Transparency	• Dependent on documentation	• Dependent on documentation	• Interested parties can view interactions at workshops • Participatory peer reviewers observe workshops, participate in Workshop #3 • Dependent on documentation	• Interested parties can view interactions at workshops • Participatory peer reviewers observe workshops, participate in Workshop #3 • Dependent on documentation
Regulatory Assurance*	• Limited or no interaction with proponent and resource experts reduces confidence • Depends on TI team and degree to which data, models, and methods are readily available	• Individual interaction with proponent and resource experts increases confidence over Level 1 • Depends on TI team; degree to which data, models, and methods are readily available; and success in obtaining additional information and understanding from individual interactions	• Interaction among proponent, resource, and evaluator experts in facilitated workshops greatly increases confidence over Level 2 • Documentation of evaluation and integration process by TI Team key to high levels of confidence	• Interaction among proponent, resource, and evaluator experts in facilitated workshops greatly increases confidence over Level 2 • Documentation of evaluation and integration process by evaluator experts key to high levels of confidence

*Regulatory Assurance is defined as confidence that views of the larger technical community has been considered and that the center, body, and range of technically defensible interpretations has been represented.

Table 4-2. Attributes of Various SSHAC Levels (Continued)

SSHAC Level	Level 1	Level 2	Level 3	Level 4
Cost	• Lowest because of limited number of participants	• Slightly greater than Level 1 because of time required for interaction with proponent and resource experts	• Significantly greater than Level 2 because of greater number of participants and use of facilitated workshops • Greater likelihood that TI team members are physically dispersed, requiring costs for systems to remotely access data and information • Costs associated with TI Team working meetings	• Comparable to Level 3 in terms of use of facilitated workshops and numbers of participants • Greater likelihood that TFI team members and expert evaluators are physically dispersed, requiring cost for systems to remotely access data and information • Greater than Level 3 because of need for TFI to interact individually with evaluator experts
Duration	• Shortest because of limited or no interaction with proponent and resource experts	• Slightly greater than Level 1 because of time required for interaction with proponent and resource experts	• Significantly greater than Level 2 because of constraints in organizing workshops around proponent and resource expert, TI team member, and PPRP member personal schedules	• Similar to Level 3 or longer because of constraints in organizing workshops around proponent, resource, evaluator expert, TFI team member, and PPRP member personal schedules
Management challenge	• Least because of greater control over participants	• Slightly greater than Level 1 because of need to interact individually with proponent and resource experts whose schedules cannot be controlled	• Significantly greater than Level 2 because of increased number of participants (a number of whom may require subcontracts) and the logistics of organizing workshops	• Greater than Level 3 because of increased number of participants (a number of whom may require subcontracts), the logistics of organizing workshops, and the logistics of organizing needed interactions among the TFI team and expert evaluators

Of course, many cases exist where the criteria are in conflict, and these will require a project-specific evaluation of the relative value of the various factors. An example is, a study for a nuclear power plant sited in a country having very little regulatory oversight, high levels of public scrutiny, high levels of technical uncertainty, and severe resource limitations in terms of the available time to conduct the study. In this case, the sponsor will need to assess the relative value of achieving an expedited schedule versus facing public opposition to the project or not achieving regulatory approval. These types of assessments of the relative value of various potentially conflicting objectives can be addressed systematically using tools advanced in the decision analysis community (e.g., Keeney, 1996).

The Standard ANSI/ANS-2.29-2008 addresses the issue of selection of the appropriate SSHAC Level[10] (Section 4.3 of the Standard). The approach suggested in that document is summarized here to provide additional insight into the factors that influence the SSHAC Level decision. The Standard assumes that the hazard study will be for a particular site and not a regional hazard study.

The Standard gives a procedure that considers three project attributes that are used to decide on one of the four Levels:

1.   Risk-significance of the facility, defined by the highest seismic design category (SDC) of the structures, systems, and components (SSCs) of the facility. The definitions of the SDCs are given in ANSI/ANS-2.26-2004, Table 1. A category is assigned to an structure, system, or component, which is a function of the severity of adverse radiological and toxicological effects of the hazards that may result from the seismic failure of the structure, system, or component on workers, the public, and the environment. SSCs may be assigned to SDCs that range from 1 to 5. For example, a conventional building whose failure may not result in any radiological or toxicological consequences is assigned to SDC 1; a safety-related structure, system, or component in a nuclear material processing facility with a large inventory of radioactive material may be placed in SDC 5.

2.   Table 4-3 specifically defines the "nominal ground motion hazard level."

Table 4-3. Selection of Nominal Ground Motion Hazard Level

Table 2 of the Standard ANSI/ANS-2.29-2008	
MCE spectral response acceleration[a]	Nominal ground motion hazard level
<0.1g 0.1 to 0.3 g > 0.3g	Low Moderate High
[a]Maximum considered earthquake (MCE) defined as the average of 0.2- and 1.0-second period spectral responses (for 5 percent damping and assuming Site Class B) from U.S. Geological Survey maps in ASCE/SEI 7-05.	

---

[10] The SSHAC Level is termed "PSHA Level" in the ANSI/ANS Standard.

3.   Level of uncertainty and controversy, either low or high. Table 4-4 shows these three "parameters" and provides "guidance" for the "recommended minimum" Level.

Table 4-4. Guidance for Selection of Minimum (PSHA) Level

Table 3 of the Standard ANSI/ANS-2.29-2008			
SDC	Nominal ground motion hazard level	Level of uncertainty and controversy	Recommended PSHA Level[a]
3	Low	Low	1
		High	1
	Moderate	Low	1
		High	2
	High	Low	2
		High	2
4	Low	Low	1
		High	2
	Moderate	Low	2
		High	2
	High	Low	2
		High	3
5	Low	Low	2
		High	3
	Moderate	Low	3
		High	4
	High	Low	3
		High	4
[a]Minimum level of PSHA permitted.			

Although the procedure given in the Standard can provide useful information, it should only be used as a guide and not as a prescription, for several reasons. First, the identified decision factors (facility type, level of hazard, and degree of uncertainty/controversy) are only a subset of the considerations that typically enter into the decision. As discussed previously, there are issues related to the regulatory assurance that come with higher SSHAC Levels, and issues related to the resources required to conduct the higher SSHAC Levels. The Standard is silent on those issues. Second, the notion that the higher the ground motion hazard level, the higher the recommended SSHAC Level is subject to debate. Increasing the SSHAC Level increases the confidence that data, models, and methods of the larger technical community have been considered and the full center, body, and range of technically defensible interpretations represented. A site within a "high" ground motion region may have a fairly well-defined hazard level because of abundant data, while a site within a low-activity stable continental region may have much less information to define the hazard. These issues would be addressed by the "high" or "low" uncertainty level, but it is not clear that there is anything that would argue inherently that a more active region should require a higher SSHAC Level. The influence of the nominal hazard level, in combination with the level of uncertainty/controversy, can be very significant, as shown in Table 4-4. For example, an SDC-5 facility (e.g., a nuclear power reactor) sited at a low nominal hazard site (such as in the eastern United States) and with low uncertainty/controversy is permitted to conduct a Level 2 study under this scheme. This is inconsistent with the NRC viewpoint. The same facility at a high nominal hazard site (such as in the western United States) and high uncertainty/controversy would be required to conduct a

Level 4 study. Another concern is that the approach given in the Standard assumes that the progression of SSHAC Levels is linear. In reality, the progression of SSHAC Levels is nonlinear, with small differences between Levels 1 and 2 and between Levels 3 and 4, and large differences between Levels 2 and 3 (see discussion in Section 3.3)

In their considerations of the lessons learned from past hazard studies, Hanks et al. (2009, p.21) makes the following recommendation:

> *"While we recognize that the choice of SSHAC level belongs to the project sponsor who will be paying for it, we recommend that this decision be made in conjunction with the regulator, so that the sponsor has a reasonable expectation that the final results will meet regulatory requirements."*

To the degree that such communication between the sponsor and the regulator will clarify the positions of both parties relative to the decision factors discussed previously, this idea is endorsed. In the end, the decision regarding the SSHAC Level rests with the project sponsor, who must weigh his/her desire to minimize costs while maximizing benefits. Those studies that involve a broad sponsor group, such as the Central and Eastern United States (CEUS) SSC and the Next Generation Attenuation Relationship for Eastern North America (NGA-East) projects, benefit from having a range of perspectives brought to the decision process. Experience has shown that the SSHAC Level decision made by a single sponsor will require some level of explanation and defense, particularly within contentious and heavily regulated environments. The explanation will help those responsible for sponsoring the study understand what they are buying and those reviewing the study understand the factors that underlie the decision.

The goal of all SSHAC Levels is the same—to consider the views of the larger technical community and to capture the center, body, and range of technically defensible interpretations. The higher SSHAC Levels increase the probability that this has occurred appropriately and increase the confidence on the part of the regulator as well. Experience during past studies and due consideration of the potential future uses of this guidance leads to the conclusion that high levels of regulatory assurance are necessary for highly safety-significant facilities such as nuclear power plants. It is therefore strongly recommended that hazard studies for new nuclear power plants and other highly safety-significant facilities be conducted at SSHAC Levels 3 or 4 to provide the necessary levels of assurance. If a site has a viable pre-existing Level 3 or 4 study, then a Level 2 process can be used to update the pre-existing study. Chapter 6 provides additional discussions of the updating process.

Figure 4-1 shows the interrelationships among the various participants and activities in a Level 3 or 4 process. As discussed in Section 3.1 and shown in Figure 4-1, the SSHAC process consists of the evaluation phase and the integration phase, both of which are followed by the documentation phase of the project. The evaluation phase is focused on technical evaluation of the available data, models, and methods from the larger technical community. During the evaluation phase, hazard-significant data are identified, compiled in the project database, and evaluated for their quality and specific relevance in the technical assessments. In addition, alternative models and methods proposed by proponents in the larger technical community are identified and evaluated relative to their consistency with available data and support within the technical community. Consistent with the activities associated with the evaluation phase of the project, the first two workshops are focused on identifying hazard significant issues and the data available to address those issues (Workshop #1) and alternative models and methods that have been proposed by the larger technical community (Workshop #2).

The integration phase of the project begins after all applicable data, models, and methods in the larger community have been evaluated. During the integration phase of the project, the evaluator experts build models (logic-trees) that capture the current knowledge and uncertainties regarding the technical inputs to the hazard analysis. The consistency of proposed models with available data is assessed and tools for quantifying uncertainties in conceptual models and parameter values are used. The goal of the integration phase is to develop integrated models that capture or represent the center, body, and range of technically defensible interpretations. Consistent with the activities associated with the integration phase of the SSHAC process, the third workshop is focused on the discussion of the preliminary integrated models (Workshop #3, see Figure 4-1). At this workshop, feedback is provided to the evaluator experts in the form of hazard calculations and sensitivity analyses that provide insight into the relative importance of various components of the preliminary models. This information is used to assist the evaluator experts in understanding the most important components of their models and the relative contribution that component uncertainties make to the total uncertainty in the preliminary integrated model. They then use that information to prioritize their efforts in completing the integration process by the construction of the final integrated models.

The documentation phase of the process consists of developing a Hazard Input Document that specifies the particular elements of the integrated models such that they can used as input to the hazard calculations. The development of the draft and final project report provides the fundamental culmination of the SSHAC process. The report includes a discussion of the process followed in the project and all technical assessments made.

Given that a Level 3 or 4 process has been selected as appropriate for a particular project the discussions in Sections 4.3 through 4.10 below provide a description of the recommended implementation of the essential steps.

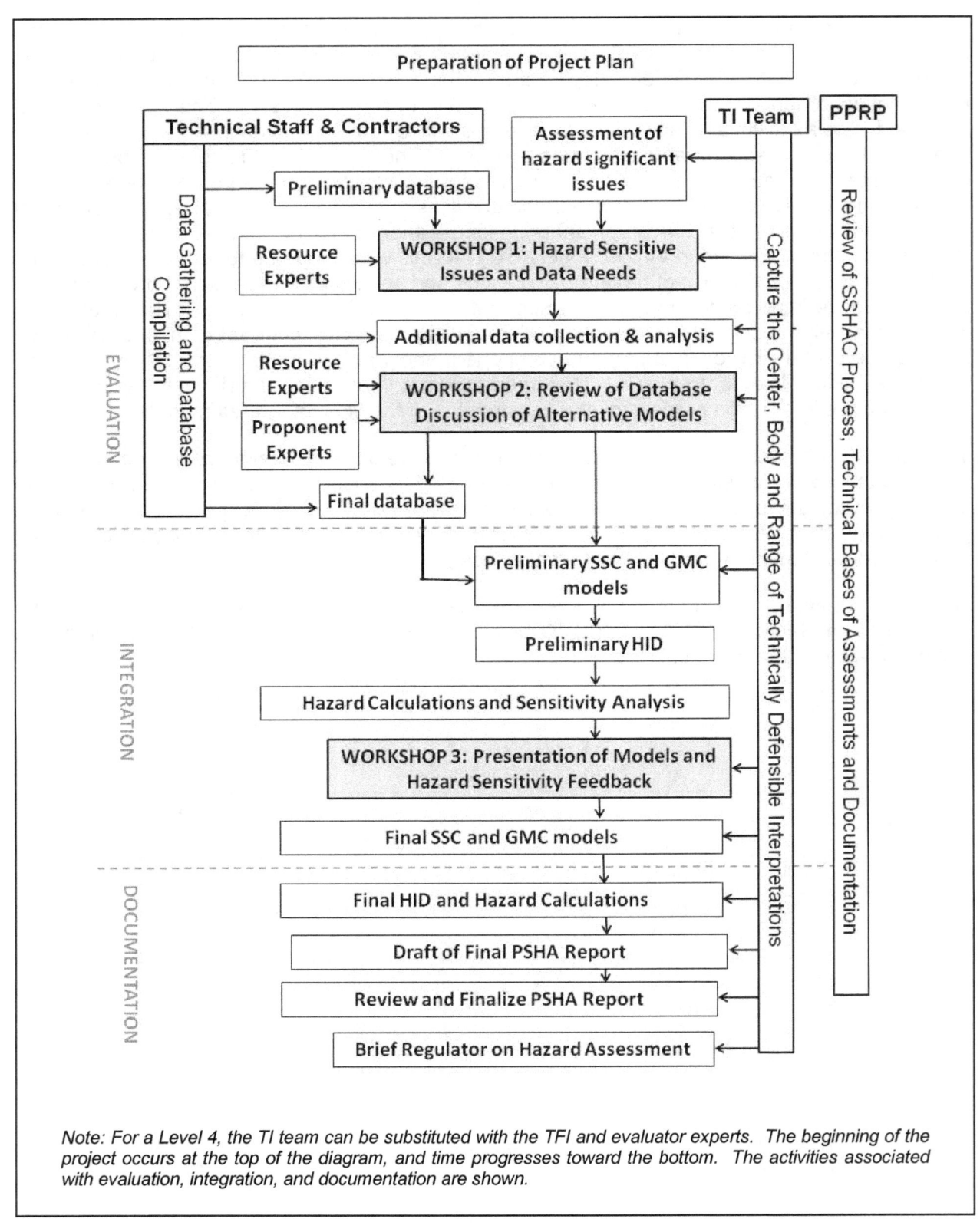

Figure 4-1. Diagram Illustrating the Various Participants and Activities that Occur Within a SSHAC Level 3 or 4 Process.

4.3 Development of Project Plan

The Project Plan is developed by the sponsor of a SSHAC project and is the fundamental tool for documenting and communicating the specific elements and details of the SSHAC Level 3 or 4 assessment process that will be used. It also provides a fundamental tool for the proper management and monitoring of a study to ensure that all procedural steps are taken and that they are conducted in a timely manner. Typically, the Project Manager and the Project TI develop the Project Plan so that all programmatic and technical activities are properly and completely described. Once the Plan has been developed and approved by the sponsor of the study, it should be considered to be "final" and subject to revision only in the unusual case of significant new information affecting the timing or resources required to conduct the study.

Acknowledging that project-specific applications may suggest that additional information be provided, the Project Plan should include the following:

1. Introduction and Context of the Study

This section should include a description of the context within which the study is being carried out including the sponsors of the study, previous hazard studies and their applicability, and significant new data or developments leading to the need to conduct the study.

2. Objectives of the Study

This section should include a description of the expected results of the study and the manner in which they will be used (e.g., design criteria, risk analyses). As applicable, the deliverables of the study should be described (e.g., types of ground-motion measures, annual frequencies of interest, time periods of interest for the facility). The regulatory framework and the manner in which the study will be used to address applicable regulations and regulatory guidance should be discussed. If applicable, the discussion should also include a description of how the study will be used to address applicable public and programmatic concerns.

3. Selection of SSHAC Level

A description of the decision process for arriving at the use of a particular SSHAC Level should be provided. The discussion should include the decision criteria that were considered and the manner in which the criteria were evaluated. Due consideration should be given to the factors discussed above in Section 4.2 in addition to other project-specific considerations deemed to be important to the sponsors. If applicable, communications with regulators regarding the SSHAC Level should be summarized.

4. Project Organization

This section should define and describe the key components of the project organizational structure including the positions, their functions, and the reporting hierarchy for the project. Section 5.2 of this report provides examples of project organizations based on recent studies. It is not necessary for the Plan to specify the names or organizations if these have not yet been selected. Each of the functions should be described in terms of their roles within a SSHAC Level 3 or 4 process, their scope of work, and the lines of communication that will be followed.

5. Work Plan Key Task Areas

This section of the Project Plan describes the principal work activities that will compose the study. Task descriptions should be given for the essential elements of a SSHAC Level 3 or 4 process, including:

- Selection of project participants: Project Manager, PTI, TI Leads, or TFI for technical issues, PPRP, TI Team members in a Level 3 or evaluator experts in a Level 4, Database Management team, Hazard Calculation team, specialty contractors, and resource experts.

- Development of project-specific database including the compilation and analysis of available data and plans for dissemination of the database to project participants.

- New data collection and analysis activities (if planned) and description of their use in addressing technical issues.

- Description of workshops including their purpose, participants, timing, and expected products.

- Model development activities including expected working meetings, their focus, duration, and expected products.

- Plans for development of hazard input documents (HID) to provide input for preliminary feedback calculations as well as to transmit the final models to the hazard analyst.

- Hazard calculations and sensitivity analyses planned to provide feedback as well as arrive at the final hazard results.

- Activities associated with developing a Draft Project Report, planned review activities, and a Final Project Report.

- Description of all planned activities for the PPRP including the manner in which the Panel will observe and review all key project activities during the course of the project, as well as review of the Draft Project Report.

6. Project Schedule

This section should provide a discussion and depiction of the timing and duration of all project activities. It is also useful to display the relationships among the project activities (e.g., the ways that a given activity requires predecessor activities to be conducted and the successor activities that will use the results of each activity).

7. Deliverables

The project deliverables should be described in sufficient detail to provide confidence that the project will meet the project objectives and realistic cost and schedule estimates can be developed. This description will also provide a basis for users of the results of the study to understand exactly what they can expect the project to deliver.

4.4 Selection of Project Participants

Given the roles that are essential to a SSHAC Level 3 or 4 process (Section 3.6) and mindful of the project-specific organization defined in the Project Plan, this activity involves identifying candidates who will fulfill the specific roles required for the project. It is useful to define a set of criteria against which candidates can be evaluated. The criteria will vary with the particular position that is being filled. For example, as discussed in Section 3.6, the specific roles and responsibilities of the PPRP lead to the definition of a particular set of attributes that must be fulfilled for a candidate to be considered. Likewise, the TI and TFI roles typically entail facilitation and management experience as well as technical expertise in the subject hazard area.

Because the technical work that is done by the TI Team in a Level 3 study and Evaluator Experts in a Level 4 study is paramount to the success of the study, it is recommended that careful consideration be given to defining explicit selection criteria based on the attributes given in Section 3.6. Guidance on selection criteria for Evaluator Experts provided in NUREG-1563 provides useful insights. This guidance states "the panel of experts selected for elicitation should comprise individuals who: (a) possess the necessary knowledge and expertise; (b) have demonstrated their ability to apply their knowledge and expertise; (c) represent a broad diversity of independent opinion and approaches for addressing the topic(s) in question; (d) are willing to be identified publicly with their judgments; and (e) are willing to identify, for the record, any potential conflicts of interest," (Kotra et al. 1996, p. 23). As another example, the selection criteria used to select the Evaluator Experts for the Yucca Mountain PVHA-U were the following:

1.   Earth scientist of high professional standing and widely recognized competence based on academic training and relevant experience. Tangible evidence of expertise, such as written documentation of research in refereed journals and reviewed reports is required.

2.   Understanding of the general problem area through experience collecting and analyzing research data for relevant volcanic studies in the southern Great Basin or similar extensional tectonic environments; prior familiarity with the data available for the Yucca Mountain site will be an asset, but not a requirement for participation.

3.   Availability and willingness to participate as a named panel member including a commitment to devoting the necessary time and effort to the project and a willingness to explain and defend technical positions.

4.   Personal attributes that include strong communication and interpersonal skills, flexibility and impartiality and the ability to simplify. Individuals will be asked specifically not to act as representatives of technical positions taken by their organizations but rather to provide their individual technical interpretations and assessments of uncertainties.

5.   Selection would contribute to a balanced panel of experts with diverse opinions, areas of technical expertise, and institutional/organizational backgrounds (e.g., from government agencies, academic institutions, and private industry). (SNL, 2008, p.218)

These types of selection criteria are applicable to the selection of both the TI Team members and the Evaluator Experts in Level 3 and 4 studies, respectively. It is recommended that a pool of candidates be identified based on general application of the criteria, followed by a selection based on a closer evaluation against the criteria. The number of TI Team members typically ranges from 5 to 15 and the number of evaluator experts from 5 to 10. The selection of the TI Team members should be made by the Project Manager and the PTI in a Level 3 study and by the Project Manager and the PTFI in a Level 4 study.

One advantage of establishing an explicit set of selection criteria—and informing the chosen individuals of those criteria—is that they provide a basis for evaluating the performance of the individuals during the course of the study and, if necessary, removing them from the study. For example, if selection criteria call for willingness to commit the necessary time to the study and they are not willing or able to do so, grounds would exist to remove them from the study. Likewise, if they are not willing or able to provide independent assessments and play the role of an impartial evaluator expert, they should not remain in that role on the project.

It may be useful to require that TI Team members or evaluator experts to disclose any potential conflicts of interest to remove any doubt that they are acting as independent evaluators and not as representatives of their agencies or under the influence of any business relationships. For example, the Yucca Mountain PVHA-U project asked each expert to disclose information that could represent a potential conflict of interest. The information included:

- Organizational affiliations and relevant business relationships.
- Sources of research support.
- Other circumstances that might be construed as creating a potential conflict of interest.

As endorsed in the original SSHAC guidance (NUREG/CR-6372, p.28), the selection criteria for the PPRP should likewise be carefully considered and made explicit. As discussed in Section 4.6.8, the roles and responsibilities of the PPRP demand that the members have expertise both in the technical issues being addressed as well as the SSHAC process being followed. If these qualities cannot be found in each individual PPRP member, then the panel as a whole should possess them. Typically, the number of PPRP members varies from 4 to 8, but fewer may be sufficient. The selection of the PPRP members should be made by the Project Manager and the PTI in a Level 3 study or the Project Manager and PTFI in a Level 4 study.

As discussed in Section 5.2, it is suggested that the organization of a Level 3 study include a PTI as well as TI Leads for each of the technical areas being considered (e.g., seismic source characterization, ground motion characterization). Also, for a Level 4 study, the TFI is usually a small team of 2 to 3 people including a Project TFI and TFIs for each of the technical areas being addressed. This will ensure adequate communication and eliminate the potential for unresolved interface issues between the technical areas. The Project Manager makes the selection of the PTI, TI Leads [Level 3], PTFI, and TFIs [Level 4] in consultation with the project sponsors. Several desired attributes of the candidates should be considered including:

- A thorough understanding of the SSHAC goals and processes.

- Technical expertise in the issues being addressed and in hazard analysis.

- Experience in conducting previous Level 3 and 4 studies (in a capacity as an evaluator expert or PPRP member).

- Strong communication skills to work with the technical evaluators.

- Project management skills to ensure technical products are high-quality and timely.

- Understanding of the regulatory framework for the study and its subsequent use.

4.5 Development of Project Databases

4.5.1 Compilation of Available Data

Compiling data for the project is critical to developing a model that is based on the most complete and up-to-date information. Documenting the effort is important for demonstrating that efforts have been made to consider the range of views of the technical community. Data compilation should begin at the time of project authorization and continue to the point that the final models are developed. This task begins with specification by the evaluator experts of the data that they think they will need. The project database is augmented based on data needs identified at Workshop #1 on Significant Issues and Available Data. This augmentation continues during the course of the evaluation process and model development. Where appropriate, data will be placed in a common GIS format that is readily usable by the project participants. Data sources should include, as appropriate available information from the following:

- Professional literature.
- Data held in the public domain by groups such as government agencies.
- Private domain data developed as part of exploration activities or other projects.
- Available data in the academic sector and other research institutions.
- Site-specific data developed in the site vicinity (for site-specific studies).

As the project progresses and assessments are made, the database management activity should include preparation of derivative maps and products that are directly applicable to the hazard analysis (e.g., seismicity maps). Additional analysis may also be necessary to provide the information and analyses that the evaluators need for their assessments. As part of the project database, a comprehensive bibliography of applicable literature should be compiled for use by the evaluators as well as to provide documentation that the evaluators had knowledge of the literature at the time of their assessments.

4.5.2 Collection of New Data

Early in the project during the data compilation stages, it may be possible to identify significant gaps in the available data that may significantly impact the hazard results. If project resources allow, focused new data collection efforts can be conducted within the timeframe of the SSHAC study to assist in the technical evaluations. Several past studies have benefited from such data collection efforts. For example early in the Yucca Mountain PVHA-U study, the evaluator experts indicated that there existed significant uncertainty in the spatial distribution and ages of possible buried volcanic features in the site region. The evidence for the possible features at the time was primarily aeromagnetic and gravity data coupled with limited borehole data. The TFI team evaluated the information and concluded that the age and spatial distribution of these potential features could have a significant effect on the hazard results. A focused drilling and sampling program was developed and implemented with the intention of directly establishing the nature of the geophysical anomalies. The new data collection program occurred within the timeframe of the hazard study and provided a substantial reduction in the uncertainties in key model inputs.

Any new data collection activities should be identified early in the project, evaluated for their potential impact on the hazard results and associated uncertainties, and completed in a timely manner for use in the technical evaluations. Typically, this would mean that the activities should be completed prior to Workshop #3 on Feedback and certainly no later than the time that the models are finalized.

67

The PTI/PTFI should take the responsibility for identifying any new data collection activities in consultation with the Project Manager, evaluator experts, and possibly the PPRP. Of course, the sponsor of the study ultimately makes the decision regarding whether or not such activities should be carried out because of the need for additional resources. If a decision is made to proceed, the PTI/PTFI and applicable TI/TFI Leads should assume responsibility for completing the studies in a timely manner and for overseeing the incorporation of the results into the project database.

4.5.3 Data Dissemination

Given that the participants of Level 3 and 4 studies are often geographically distributed, a key part of the database management activity will entail the dissemination of the database in an efficient manner for use by all. Various alternatives exist for the data dissemination including Web sites, secure portals, ftp sites, on-request distribution, etc. For example, the PEGASOS project established a secure Web-based portal that allowed the project participants to access the project database from any location. The Thyspunt PSHA project has coupled a secure portal with a searchable online reference database. Regardless of the mechanism chosen, it is imperative that the evaluator experts have ready access to the data for their evaluations and model building at any time. As will be discussed in Section 4.6 and 4.7.1, the real "work" done by the evaluators occurs between the workshops. Multiple "working meetings" are held so that the evaluator teams can review the data and make their assessments. For this to work efficiently, the Database Management team should interact with the TI/TFI Leads to develop a database structure and index that is appropriate and intuitive for the users. In addition, the Database team should be prepared to develop any derivative data products that are requested such as combinations of GIS data layers or sorting of the data according to attributes specified by the evaluators. Often, this will need to be conducted "real-time" during the working meetings to inform the evaluation process.

4.6 Workshops

Workshops play a vital role in SSHAC Level 3 and 4 processes. They provide opportunities for key interactions to occur; for models and interpretations to be presented, debated, and defended; and for sponsors and reviewers to observe the progress being made on the study. As will be noted further below, however, they are *not* the place where models are developed and technical assessments are made. That will occur between the workshops.

The hallmark of a SSHAC Level 3 or 4 process is the interaction that occurs in a series of structured, facilitated workshops. Each workshop has a specific focus and goal, and each requires that particular work activities have been conducted prior to its occurrence and certain work activities will occur following. Typically, each workshop entails 2 to 4 days. The ground rules for the workshops need to be established and presented at the beginning of each workshop so that the attendees understand and can fulfill their particular roles. The ground rules also need to be consistently enforced. Although additional workshops and gatherings (e.g., field trips) can be conducted for any particular project, Sections 4.6.1 through 4.6.3 discuss the required workshops and their focus.

4.6.1 Workshop #1 Significant Issues and Available Data

The goals of workshop #1 are (1) to identify the technical issues of highest significance to the hazard analysis and (2) to identify the available data and information that will be needed to address those issues. Because this is the first of a series of workshops, it is valuable to begin the workshop with a summary of the entire project and its objectives. This should include the expected deliverables and pertinent information related thereto (e.g., the range of annual frequencies of exceedance of interest, the timeframe of interest, etc.).

Workshop #1 is the first opportunity for the TI/TFI Leads to establish the ground rules for all of the workshops. The workshop should begin with a clear definition of the goals of the workshop, an explanation for the process that will be followed, and a definition of the roles of all those who attend.

Typically, those in attendance at the workshops are the following participants:

- Project sponsors.
- Project Manager.
- PPRP.
- Project TI [Level 3] or Project TFI [Level 4].
- TI Teams/Evaluator Experts including the TI Leads/TFI Leads.
- Database Manager.
- Specialty Contractors, as appropriate.
- Regulators.
- Resource experts (expertise varies with each workshop).

First and foremost, the workshops are held to provide information to assist the TI/TFI Leads and the evaluator experts in their technical assessments. Therefore, it is essential that they are given ample opportunity to ask questions and to thoroughly understand what is being presented. All other workshop attendees are "observers" except as their participation is required. For example, at the first workshop, resource experts are asked to present and discuss their databases. Questions following each presentation should first come from the evaluator experts, not from other resource experts or other workshop attendees. The TI/TFI Leads are responsible for ensuring that the agenda is kept and all presentations are allowed to occur in an equitable manner. It is suggested that a short period of time to be set aside at the end of each day for any of the observers at the workshop including the PPRP to make statements or to pose questions.

The workshop ground rules should be repeated at the beginning of all subsequent workshops because the workshop attendees will change with each workshop.

Prior to Workshop #1, the TI/TFI Leads should conduct preliminary hazard studies and sensitivity studies to assist in identifying hazard-significant issues based on available data. The sensitivity analyses should be presented and discussed at the workshop, and they can be supplemented with considerations of hazard sensitivity at other sites and the issues that have generally been shown to be important, based on experience. The purpose of identifying the hazard-significant issues first is to provide a basis for focusing and prioritizing the database development efforts. Experience has shown that technical experts are usually fascinated with discussions of technical data, but they often are not well informed regarding the issues that are most important to a hazard analysis. This part of the workshop should end with a listing of the hazard-significant issues and a listing of the types of data that can best address the issues.

The second part of Workshop #1 should focus on the data that are available to address the hazard-significant issues identified in the first part of the workshop. The discussions of the available data should be through a series of presentations by resource experts who have developed specific datasets. For example, for a seismic hazard analysis, the resource experts could discuss available seismicity catalogs, studies of historical earthquakes, regional and local geophysical data, geologic studies of tectonics, ground-motion recordings, geotechnical site data, and the like. These data resource experts should come from a wide range of affiliations and aspects of the technical community. The goal of this part of the workshop is to assist the TI/EE members in identifying the data and information that should be made part of the project database and in understanding the attributes of the available datasets (e.g., the precision, drawbacks, etc.) to the extent possible. Therefore, it is important to instruct the resource experts to discuss not only the data but the accessibility of the data for use on the project. For example, are the data publicly available? How can they be accessed as reports, publications, etc.? The experts can also each be asked to provide their knowledge of additional available data—beyond that given in their presentations—which the project should also consider. For example, in the CEUS SSC project, the resource experts attending the workshop were asked to provide reference lists of data that they were familiar with following the workshop. During the course of the data identification process, the evaluator experts may identify data gaps that can be filled with new data collection efforts within the time and resource constraints of the project. If data collection activities are contemplated or have been initiated at the time of the workshop, they should be described in detail at this workshop.

One concern that often emerges in the first workshop is the tendency for the resource experts to move from merely a presentation of available data into discussions of their interpretations of the data and the models that they have developed from it. In most earth science problems, no clear-cut boundary exists between what we call "data" and what we call "interpretations." For example, seismic reflection "data" can mean the profiles devoid of any interpretation of reflectors, or to include the interpretation of reflectors (faults, folds, beds, etc.), but without interpretations of multiple profiles in a three-dimensional manner. The point here, from the standpoint of the SSHAC process, is that evaluator experts do evaluations of the data for purposes of the hazard analysis. Moreover, the forum for hearing and debating alternative interpretations of the data is Workshop #2. So the TI/TFI Leads should make every effort to limit the discussions at Workshop #1 to the available data, acknowledging that some discussion of data interpretations will inevitably occur.

4.6.2 Workshop #2 Alternative Interpretations

The goals of Workshop #2 Alternative Interpretations are (1) to present, discuss, and debate alternative viewpoints regarding key technical issues; (2) to identify the technical bases for the alternative hypotheses and to discuss the associated uncertainties; and (3) to provide a basis for the subsequent development of preliminary hazard models that consider these alternative viewpoints. The workshop also provides an opportunity to review the progress being made on the database development and to elicit additional input, as needed, regarding this activity.

A key attribute of this workshop is the discussion and debate of the merits of alternative models and viewpoints regarding key technical issues. Proponents and resource experts should present their interpretations and the data supporting them. Presentations of alternative viewpoints on the same topic should be juxtaposed, if possible, and facilitated discussion should occur with a focus on implications of the inputs to the hazard analysis (not just on scientific viability) and on uncertainties (e.g., what conceptual models would capture the range of interpretations and the relative credibility of the alternatives). Because not all proponents of alternative viewpoints may be able to attend the workshop, interpretations made by individuals who may not be present should be identified and discussed. This will help assure that all viewpoints are ultimately considered. If feasible, the TI/TFI should present those viewpoints at the workshop so that an opportunity exists to present, challenge, and defend them.

In the spirit of capturing the spectrum of thinking across the entire technical community, a goal of this workshop is to provide an effective forum for the exchange of ideas. More importantly for the SSHAC process, the workshop provides a unique opportunity for the evaluator experts to begin their consideration of the range of models and methods held by the larger technical community. Therefore, they should strive to not only understand the alternative interpretations, but also the degree to which each hypothesis is supported by the available data. The proponent experts should be asked to be prepared to discuss the uncertainties in their interpretations, the strengths and weaknesses in their arguments, and their view of the degree of support that their interpretations have within the larger technical community. It is often useful for presenters to be provided with a list of questions developed by the TI/TFI to focus the presentation on areas of interest to model development. A role of the TI/TFI should be to provide support, as needed, to proponents to ensure that interpretations are not judged on the basis of presentation skills. Proponent experts should be encouraged to interact among themselves within the structure facilitated by the TFI/TI Lead. For example, as part of discussions of the seismic potential of the New Madrid seismic zone at Workshop #2 in the CEUS SSC project, a facilitated discussion was conducted among proponents having very different models regarding the future earthquake potential of the zone. This experience showed that asking the proponents to consider the views of others in the community can encourage useful discussion.

4.6.3 Workshop #3 Feedback

As will be discussed in Section 4.7, following Workshop #2, the evaluator experts develop their preliminary models. Based on these models, preliminary calculations, and sensitivity analyses are conducted. The goal of Workshop #3 Feedback is to present and discuss the preliminary models and calculations in a forum that provides the opportunity for feedback to the evaluators. Feedback is given in the form of hazard results and sensitivity analyses to shed light on the most important technical issues. Feedback is also provided at this workshop by participation of the PPRP and allowing them to ask questions regarding the preliminary SSC and GMC models. The feedback provided at this workshop will ensure that no significant issues have been overlooked and will allow the evaluators to understand the relative importance of their models, uncertainties, and assessments of weights. This information will provide a basis for the finalization of the models following the workshop.

The workshop consists of two parts: (1) the evaluators presenting their preliminary models with particular emphasis on the manner in which alternative viewpoints and uncertainties have been incorporated and (2) sensitivity analyses and hazard calculations that provide insight into the preliminary models. In the discussions of the preliminary models, the technical bases for the assessments and weights should be described to allow for a discussion of the implications and constraints provided by the available data. This part of the workshop differs somewhat between a Level 3 and a Level 4 study. In a Level 3 study, the entire TI Team will have been involved in the development of the preliminary model, and it is not expected that individual members of the team will question aspects of the model. Rather, the PPRP will be expected to question and probe aspects of the preliminary model to understand the manner in which the views of the larger technical community have been considered and the range of technically defensible interpretations included. In a Level 4 study, each evaluator expert will present his/her preliminary model and should discuss and defend it under questioning of colleagues on the panel. Again, the questions should probe how each expert considered the views of the larger community and the manner in which their preliminary model represents current knowledge and uncertainties. If the model input distributions developed by the evaluator experts are narrow with little overlap among the other experts' distributions, this should spark some discussion regarding the bases for the assessments so that it is clear what is causing the differences. However, there is no requirement that any expert's assessment be changed as a result of the feedback discussions.

In the second part of the workshop, the presentation of the sensitivity analyses and preliminary hazard calculations provide a means of focusing the discussions on those issues having the greatest hazard significance, including the largest contributors to uncertainty. In turn, this will serve to focus the assessments performed after the workshop on those technical issues of most importance to the hazard results. Section 4.8 discusses the types of sensitivity analyses and hazard calculations that should be considered. It is important to include not only hazard calculations and associated sensitivity analyses but also sensitivity analyses that will provide insight into the models themselves. For example, the effect of various components (branches of the logic-tree) of the SSC model on the assessments of maximum earthquake magnitudes and earthquake recurrence rates could be examined. Likewise, the relative contribution that the epistemic uncertainty and aleatory variability in a particular element of the model has to an intermediate output can be explored. It should be noted that these feedback calculations are not intended to provide a basis for artificially truncating or otherwise limiting the models developed by the evaluators. Rather, they are intended to provide a basis for prioritizing the activities associated with developing the final models.

Developing and using feedback during the integration (model-building) process is an important characteristic of a SSHAC Level 3 or 4 process. In large regional studies where there are large numbers of assessments and possibly complex components to the models, the project may benefit from more than one feedback cycle. Multiple feedback cycles may be particularly beneficial in large regional studies.

4.7   Development of Models and Hazard Input Documents

Although the workshops are a hallmark of a SSHAC Level 3 and 4 process, the bulk of the expert model development process is done between the workshops. The integration or model development activity conducted by the evaluator experts includes the evaluation of available data; consideration of alternative data, models, and methods; and appropriate quantification of uncertainties. These types of technical evaluations are conducted in a nonpublic setting using typical scientific assessment processes. It should be noted that the discussion in the remainder of this chapter is based on a project that has only the three required workshops, but the information can be easily adjusted to accommodate additional workshops.

4.7.1   Model Development in Working Meetings

The recommended manner in which the technical evaluations take place in working meetings differs slightly for Level 3 and 4 projects; thus, they are described separately below.

SSHAC Level 3 Model Development

Much of the work conducted by the TI Team is carried out in multiple "working meetings" (typically, at least three) of the team. Each meeting usually lasts multiple days, and team members should be apprised of the purpose of each meeting early thereby allowing ample preparations to be made by all team members so that they are able to participate constructively. The Team Lead convenes the working meeting for that particular technical topic (e.g., SSC or GMC), and the meeting is held in a conference room environment. It is important that ample real-time electronic access to the project databases be arranged to facilitate discussion.

The first working meeting should review the purpose of the project, the context of the evaluations being made, time schedules, and deliverables. The TI Lead should reaffirm complete commitment of all team members to devote the required effort for successfully carrying out the project. This meeting also provides the opportunity to review methods for addressing uncertainties and to review the issues related to the use of expert judgment including cognitive biases (e.g., see Section 5.11). At that meeting, the team members should begin the identification of the hazard-significant issues and the databases that can be used to address those issues. The structure, format, and accessibility to the project database are other topics for discussion. This is also the opportunity to review and specify any new data collection activities that will be conducted to supplement the available data. Finally, the team should identify the resource experts to be invited to the first workshop to present their databases. The meeting should be organized early enough to provide the PPRP the list of proposed experts for their review and to schedule the experts who typically have very full schedules.

Following Workshop #1 Significant Issues and Available Data, one or more working meetings should be held to prepare for Workshop #2 Alternative Interpretations. This should include a review and discussion of the technical hypotheses that have been proposed by the larger technical community and proponents to be invited to present their interpretations at the workshop. The ongoing development of the project database should also be a topic of discussion.

Following Workshop #2 and prior to Workshop #3 Feedback, multiple working meetings will be necessary to develop a preliminary model that can be used for purposes of sensitivity analyses to provide the necessary feedback to the TI Team. During these meetings, the project database will need to be available in whatever formats the team finds most useful. For example, a database management team member should be present to respond to the teams' requests to superimpose various GIS-based maps and three-dimensional data. Team members will also want to consider the credibility of alternative hypotheses in light of the available data including any data collected specifically for purposes of the project. The TI Lead should lead the discussions and work with the team to develop an overall framework for the evaluations (often expressed as a master logic-tree) and the detailed evaluations of the relative weights on alternatives and uncertainties in associated parameters.

To develop the feedback required for Workshop #3, a full preliminary model will need to be developed, including the preliminary assessment of alternative weights and quantification of parameter uncertainties. This model will then be used for hazard calculations and sensitivity analyses that are specifically designed to provide information on the relative importance of various technical components of the preliminary model. For example, if two competing alternative models exist within the technical community, the preliminary model should include the alternatives. The feedback will illustrate the importance of their implications to the hazard results, both to the mean hazard result and the uncertainties. It should be emphasized to the TI team that the preliminary model is developed for the purposes of evaluating sensitivities and identifying significant issues. The team should not become anchored on this model after considering feedback, but should feel free to make whatever revisions are judged to be appropriate in subsequent parts of the project. If significant revisions are made, another round of feedback may be required.

Following Workshop #3, at which the preliminary results will be presented, one or more working meetings should be conducted to develop the final model. The results of the feedback provided at the workshop provide a basis for prioritizing the efforts that will need to be made in the evaluation process. The prioritization will be towards those technical issues and inputs to which the hazard results are most sensitive and toward the uncertainties in the model that contribute most to the uncertainties in the hazard results. The work will be coordinated and facilitated by the TI Lead, who will also monitor its progress. Issues of concern related to the progress being made or to interface issues between various components of the project (e.g., between SSC and GMC) will be reported to the PTI, who will be responsible for seeking solutions. During the model-building process, the TI Team may divide the work among subgroups to expedite the evaluation process. However, the full team should thoroughly review, understand, and endorse the decisions made by any subset of the team because the entire team will be expected to assume ownership of the final model.

Upon completion of the final model, the TI Team will move into the documentation phase of the project. One or more working meetings may be useful to establish the outline for the project report, to agree upon writing assignments, and to monitor the progress of the report-writing effort.

SSHAC Level 4 Model Development

The model development activities for a SSHAC Level 4 process are very similar to those of a Level 3 in their content, sequence, and relationship to the intervening workshops. However, the participants at the working meetings are different. Each Level 4 working meeting is attended by the TFI team and an individual evaluator expert (or evaluator team, if a team is being used). The evaluator experts (EEs) each develop their own preliminary and final models. The suite of models developed by the EEs is then integrated by the TFI.[11] The TFI is responsible for ensuring that all the EEs understand and have access to the project databases. The TFI serves the same function as the TI Lead in ensuring a common understanding of the issues associated with the use of expert judgment, assisting the experts on methods of quantifying uncertainties, facilitating the discussions, and ensuring that all of the model components have been addressed. Because only the EE (or EE team) is doing the evaluations in each working meeting, each meeting typically entails one day per EE. The TFI must work to ensure that the same information is disseminated across all EEs on the panel, and any issues that arise of general concern to the entire EE panel must be addressed and resolved across the entire panel. Clearly, because the number of working meetings in a Level 4 study is greater than that for Level 3, the meetings must be carefully planned and expeditiously scheduled to avoid protracting the time needed between workshops and to finalize and document the final report.

4.7.2 Hazard Input Documents

The hazard input document (HID) is the vehicle for summarizing the results of the technical evaluations made in the SSHAC process and transmitting the information relevant to the quantitative model to the hazard analysts for calculation. The idea of an HID was developed in the PEGASOS project and came from the need to have a documented mechanism to ensure that the expert models have been faithfully and accurately transmitted to the calculation team. The HID is a succinct summary of what is in the logic-tree for the preliminary and final models. It gives nodes, branches that represent the alternative models and parameters, and weights on the tree but does not provide any discussion or justification for the values. Explanation is given of any logic or instructions that are needed for the calculations. For example, alternative geometries may be depicted in an illustration with reference to a data file that specifies the geometry. The HID will contain other types of data files such as those that describe continuous distributions. In addition, special instructions or explanations for the hazard analyst can be included to ensure that the model is clearly understood. For example, the data file containing the results of two-dimensional smoothing of spatial rate density can be explained to ensure that the proper discretization is used and that the spatial distribution is properly accounted for. The

---

[11] This differs from a Level 3 assessment in which the TI team works in a group to develop the preliminary and final models.

74

HID should be complete and clear enough to ensure that the hazard analyst does not have to interpret any ambiguous information.

To ensure that the technical models developed from the SSHAC process are accurately transmitted to the hazard analyst, it is recommended that the TI or EE members review the HID and endorse its accuracy prior to its delivery to the hazard analyst. One advantage of the HID is that it also provides an accurate input to third-party 'spot-checks" of the calculation procedure. The HID has been shown to be an effective mechanism for providing an auditable document that ensures the accurate transfer of the SSC and GMC models to the hazard analyst.

4.8 Preliminary Hazard Calculations and Sensitivity Analyses

A distinguishing attribute of a SSHAC process is the interaction and learning that takes place. A key element that contributes to learning is feedback, which is information that provides a perspective to the evaluators that can assist them in their subsequent evaluations. Feedback occurs in many forms throughout the project, such as the information that is received when TI team members challenge the proponents in Workshop #2 to defend their interpretations or when the EE members present and defend their preliminary models to their peers at Workshop #3 in a Level 4 study. This information allows the evaluators to further consider their assessments in terms of whether they have properly and completely accounted for uncertainties and whether they have been successful at capturing the views of the larger technical community. As discussed in Section 3.5, this feedback, interaction, and learning process is distinctly different from the classical expert elicitation process, which assumes that experts come to the project equipped to answer the questions required for the project, and that the answers simply need to be extracted through a series of clever questions and elicitation techniques.

Following Workshop #2 and after the evaluators have developed their preliminary models, feedback is developed that includes preliminary hazard calculations and sensitivity analyses based on those models. The sensitivity analyses should be designed to show the importance of all elements of the preliminary model, including all branches of the logic-tree and their associated weights. This often means that the sensitivity shown can be not only in terms of the importance to the calculated hazard but also to intermediate results as well. For a PSHA, the sensitivity of elements of the logic-tree to calculated recurrence rates or to maximum magnitude can be identified, for example.

Figures 4-2 through 4-9 show examples of different types of sensitivity analyses used for feedback. Figure 4-2 illustrates the differences in predicted spatial recurrence-rate densities for different values of smoothing distance in a kernel approach used in the PVHA-U for Yucca Mountain. The results provide information to the expert on how sensitive the predicted rate densities are to this assessment. In a Level 4 study, it is useful for the EEs to see not only the implications of their own assessments but also those of the other members of the panel. Figure 4-3 gives an example of comparisons of recurrence rates for the PEGASOS project. In addition to implications to mean estimates, it is important for the sensitivity analyses to explore the implications of the preliminary model to uncertainty. Figure 4-4 displays the uncertainties associated with the preliminary models of the various PVHA-U experts, and Figure 4-5 shows the uncertainties derived from the ground-motion experts on the PEGASOS project. The goal of these types of comparisons is not to gain any type of consensus across the panel; rather, they are intended to trigger an examination of the reasons for the differences in terms of the technical assessments. Figures 4-6 and 4-7 show effective ways to further examine the uncertainties. These plots show the relative contribution that the uncertainties in individual components of the models (e.g., the range of values associated with the branches of the logic-tree at a particular node) make to the total uncertainty in the model. This type of information can help to inform the TI or EE experts in their subsequent assessments of the relative impact that various components of the model have on the uncertainty.

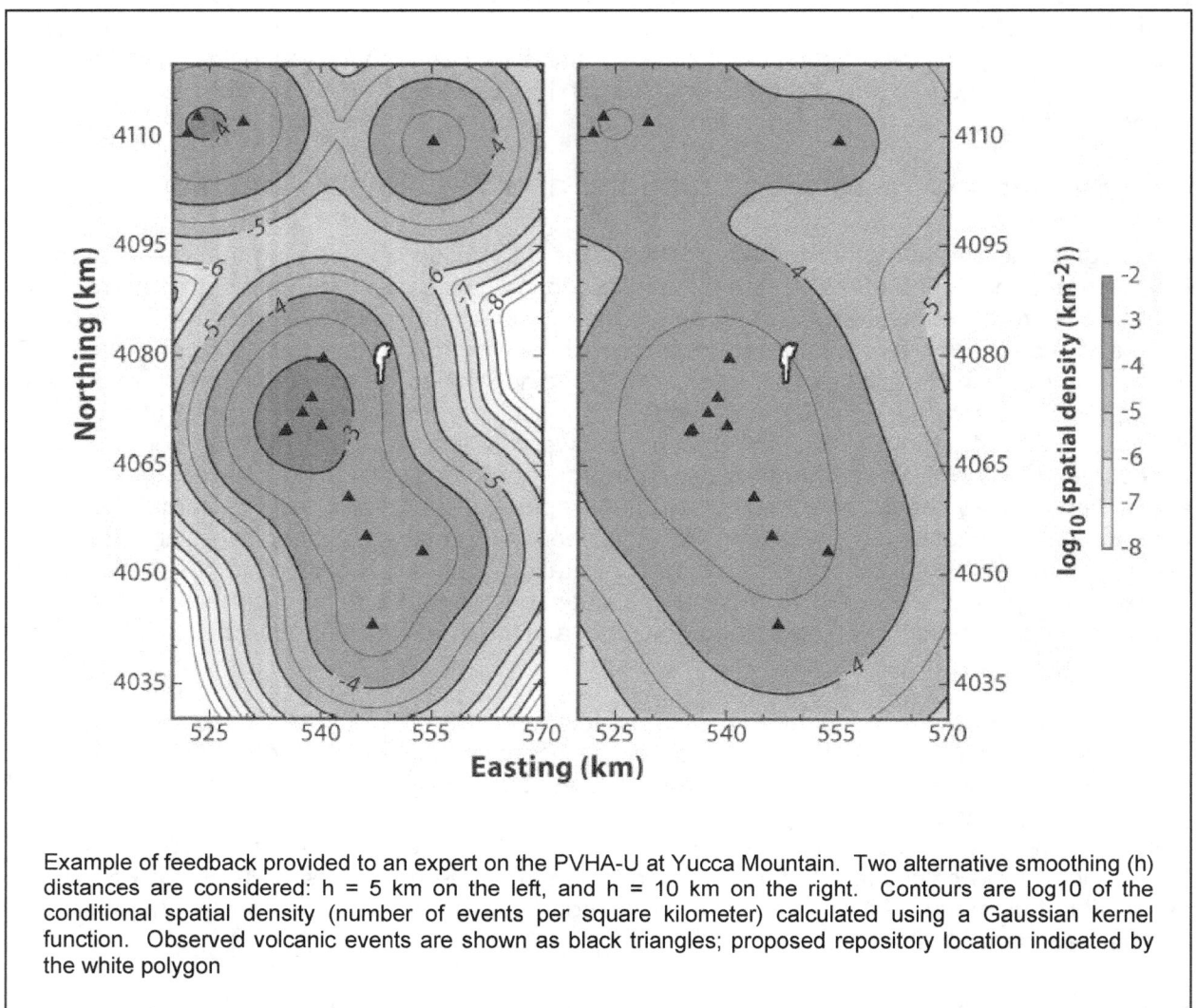

Example of feedback provided to an expert on the PVHA-U at Yucca Mountain. Two alternative smoothing (h) distances are considered: h = 5 km on the left, and h = 10 km on the right. Contours are log10 of the conditional spatial density (number of events per square kilometer) calculated using a Gaussian kernel function. Observed volcanic events are shown as black triangles; proposed repository location indicated by the white polygon

Figure 4-2. Example of Feedback on Smoothing Provided to an Expert on the PVHA-U at Yucca Mountain.

At the hazard level, sensitivity can be displayed in a number of established and useful ways. Most common are disaggregation plots that show the contribution to the mean hazard of different distance, magnitude, and epsilon[12] bins (Figure 4-8) and the plots showing the mean hazard contribution from various seismic sources (Figure 4-9). Also useful are plots of mean hazard curves that display the hazard results conditional on particular branches of the logic-tree.

---

[12] Epsilon is defined as the number of standard deviations in the ground motion model.

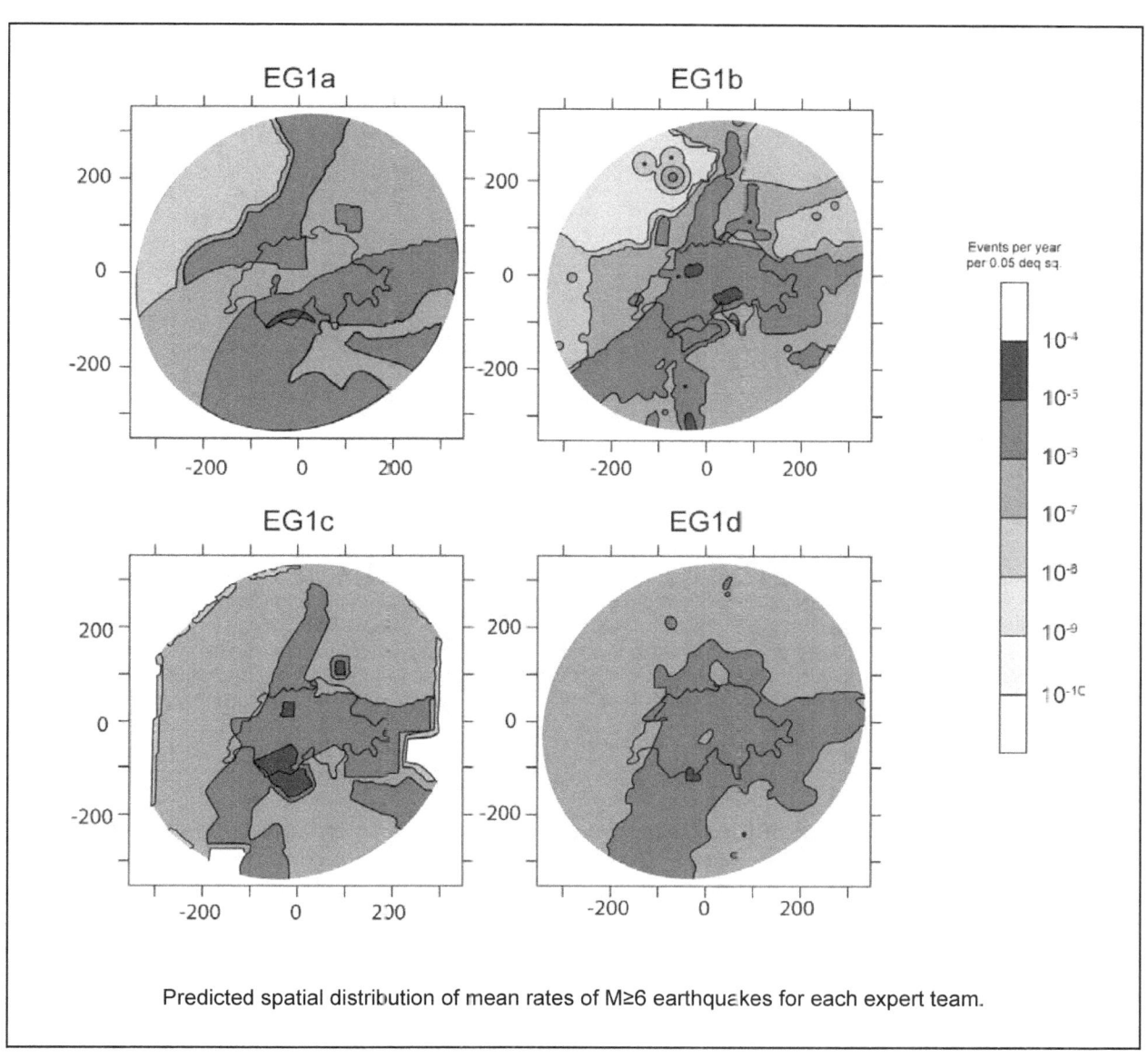

Predicted spatial distribution of mean rates of M≥6 earthquakes for each expert team.

Figure 4-3. Example of Feedback on Seismic Rates Provided to the Four SSC Evaluator Expert Teams on the PEGASOS Project, Switzerland.

In sum, feedback is needed so that the TI/EE experts can make hazard-informed decisions, and they can refine their models to focus on those elements that matter. One possible downside to providing quantitative hazard feedback is that it can allow motivational biases to occur. That is, if an evaluator is motivated to want a 'conservative" hazard estimate or one that meets some predetermined design level (e.g., the ground motion level for a specific certified design), having a knowledge of the absolute hazard level may be distracting On the other hand, the absolute amplitude of the hazard can in some cases provide useful and important information needed to finalize a model. Therefore, we recommend that the sensitivity analyses developed early in the project for Workshop #1 present hazard results in the form of relative contributions or sensitivities, normalized so that the actual hazard estimates are not shown (see Section 5.11). This provides the same valuable feedback information but would discourage motivational biases

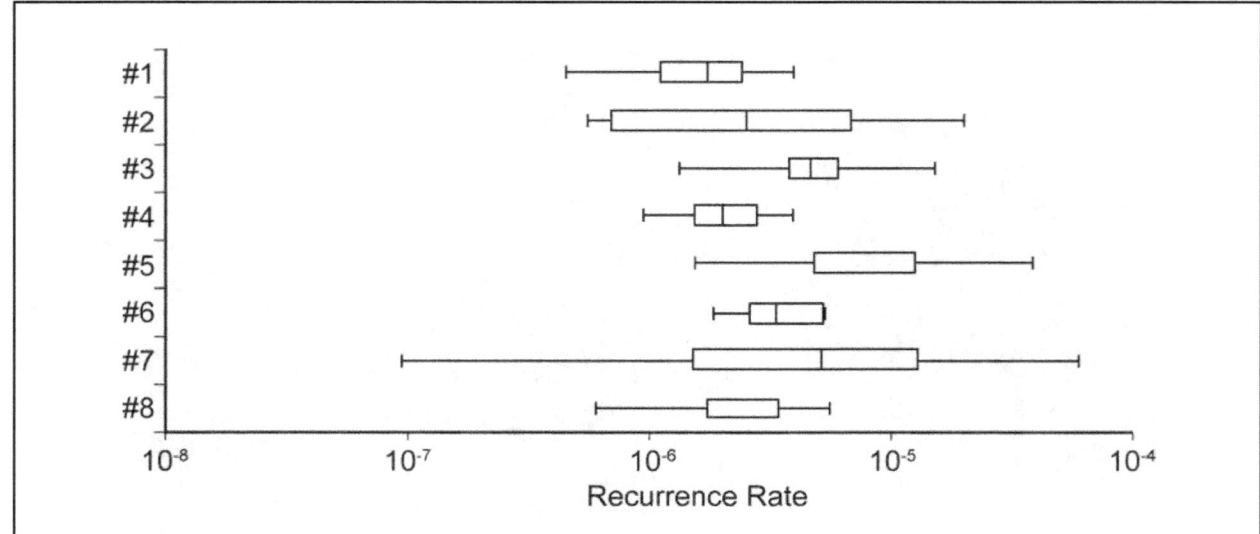

The Y-axis crosses the X-axis at the "most likely" or nominal value, which represents the annual probability of dike intersection at the repository with all model inputs set at their 50th percentile or most likely value. The length of the top bar represents the 5th to 95th percentiles of the full distribution on the probability of intersection across all model inputs. Subsequent bars illustrate the uncertainty that results from uncertainty in a single specified input. These are calculated by setting all parameters equal to their most likely values and then varying one from its lowest to highest value. This type of plot helps the expert to understand the relative contribution that uncertainties in various elements of their models makes to the total uncertainty in the hazard assessment.

Figure 4-4. Example of Feedback on Recurrence Rates and Uncertainties Provided to the Experts on the PVHA-U Project for Yucca Mountain.

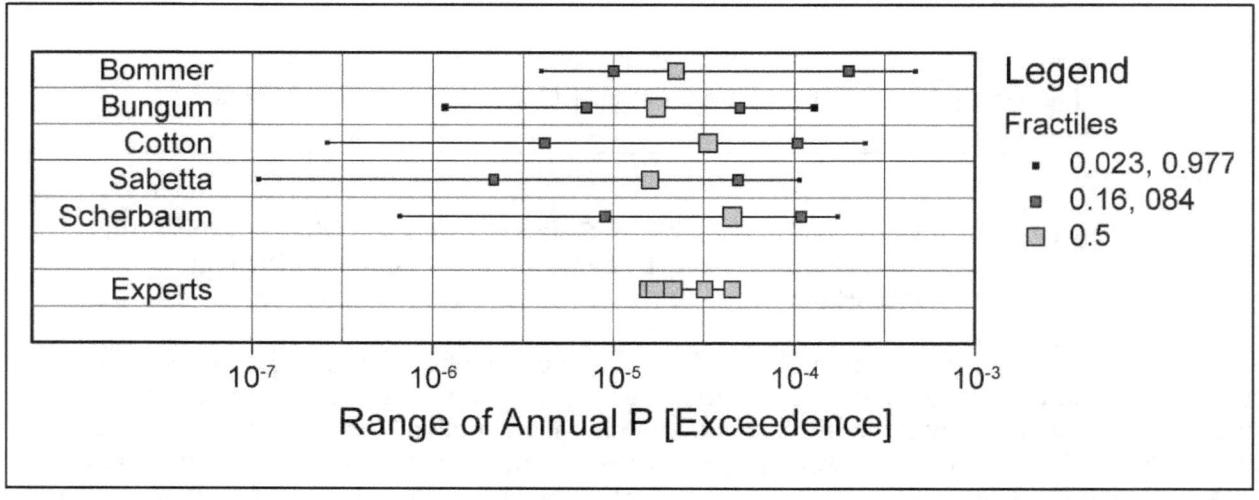

Figure 4-5. Example of Sensitivity Analysis for the Five GM Experts on the PEGASOS Project, Switzerland.

early in the project. For sensitivity analyses developed based on the preliminary models for Workshop #3, it is recommended that actual hazard results be shown because this can provide additional information needed by the evaluators to finalize their models.

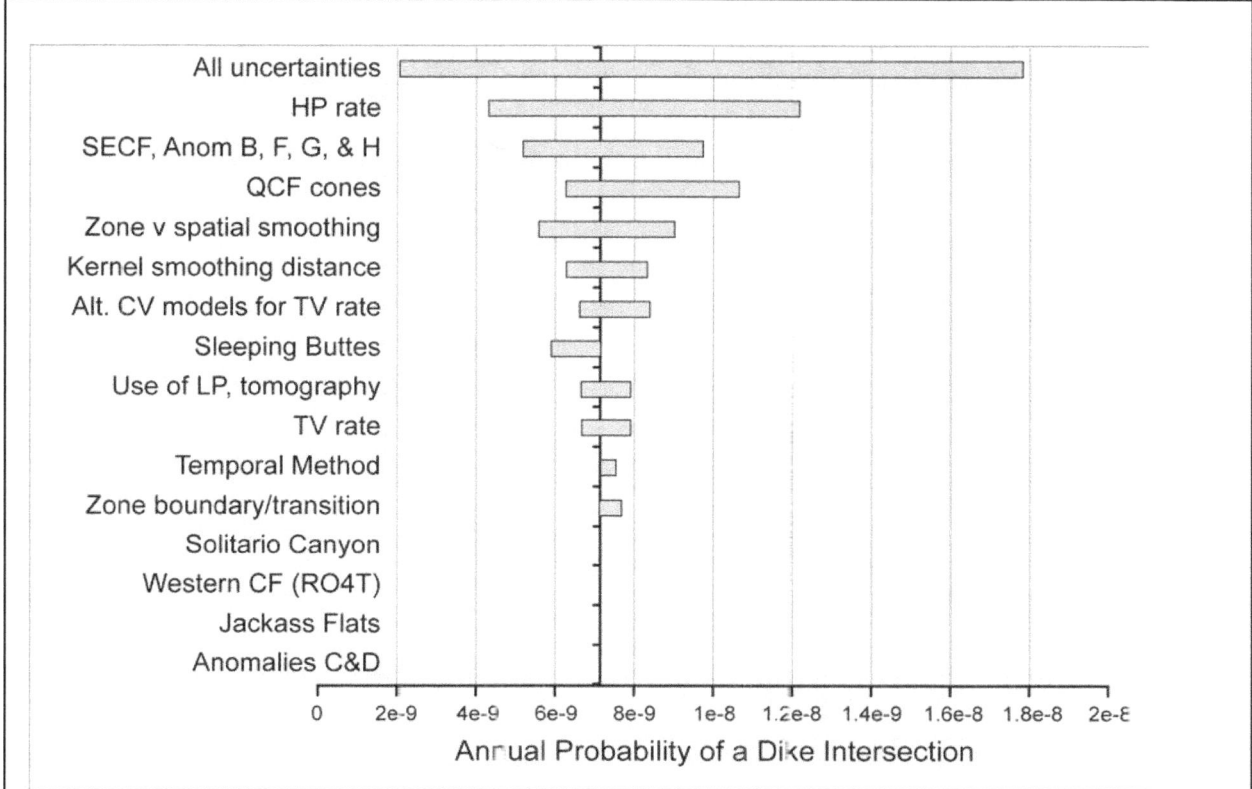

The Y–axis crosses the X–axis at the "most likely" or nominal value, which represents the annual probability of dike intersection at the repository with all model inputs set at their 50th percentile or most likely value. The length of the top bar represents the 5th to 95th percentiles of the full distribution on the probability of intersection across all model inputs. Subsequent bars illustrate the uncertainty that results from uncertainty in a single specified input. These are calculated by setting all parameters equal to their most likely values, and then varying one from its lowest to highest value. This type of plot helps the expert to understand the relative contribution that uncertainties in various elements of their models makes to the total uncertainty in the hazard assessment.

Figure 4-6. Example of Feedback on Annual Probability and Uncertainties Provided to Experts on the PVHA-U for Yucca Mountain.

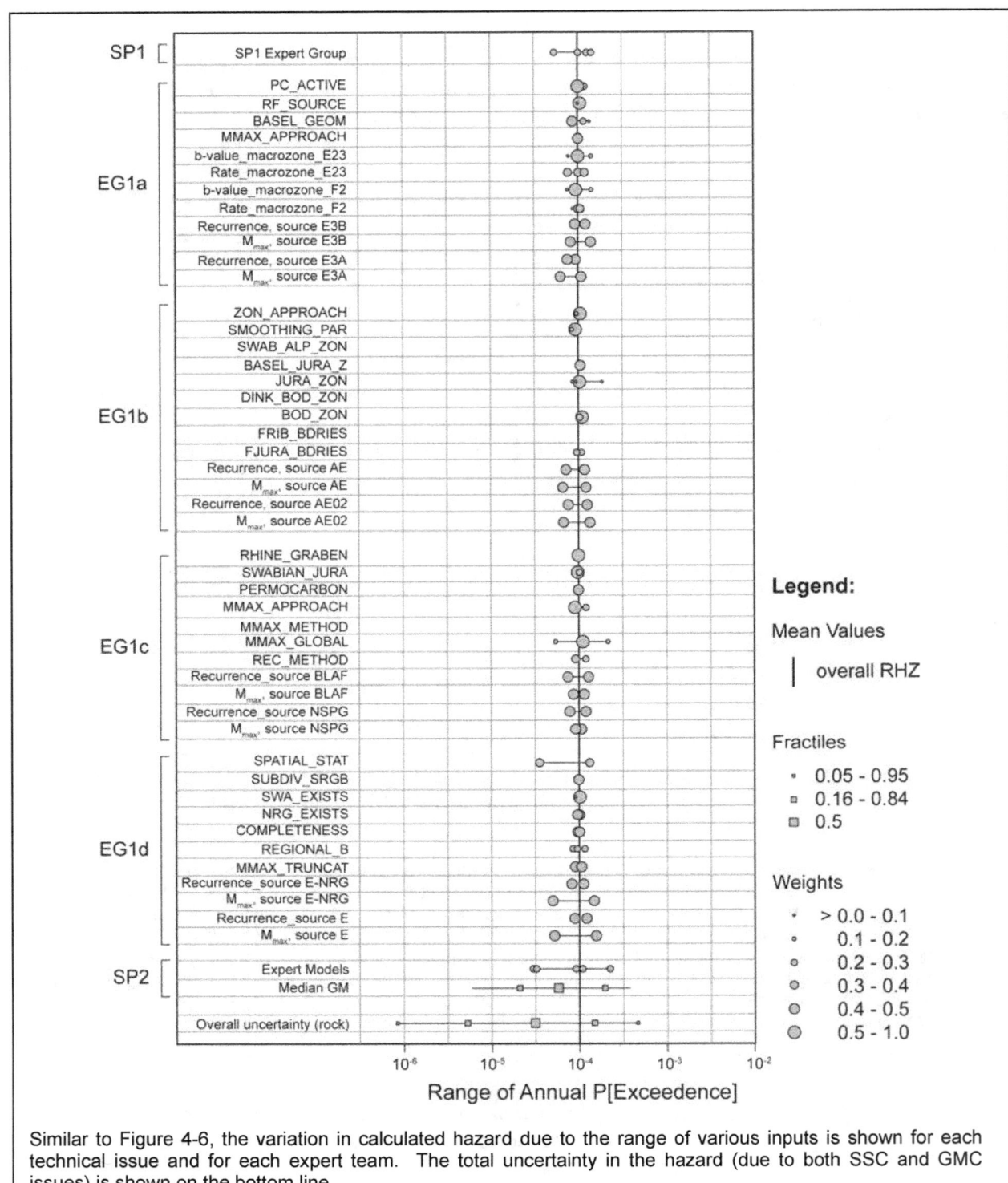

Figure 4-7. Example of a Plot Used in the PEGASOS Project to Illustrate the Relative Contribution of Various SSC Issues to the Total Uncertainties.

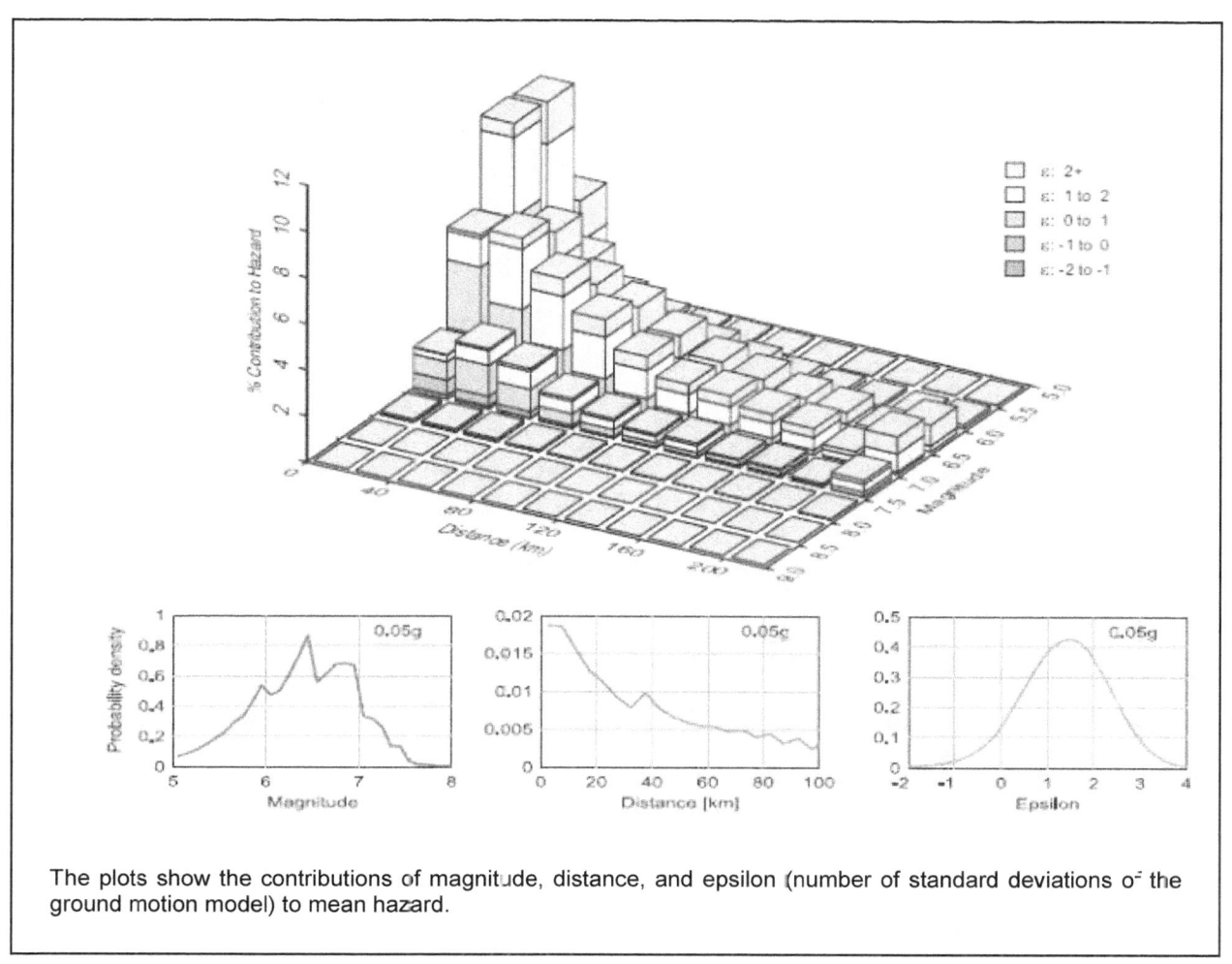

The plots show the contributions of magnitude, distance, and epsilon (number of standard deviations of the ground motion model) to mean hazard.

Figure 4-8. **Example of Disaggregation Results for PSHA Results at a Site for the PEGASOS Project.**

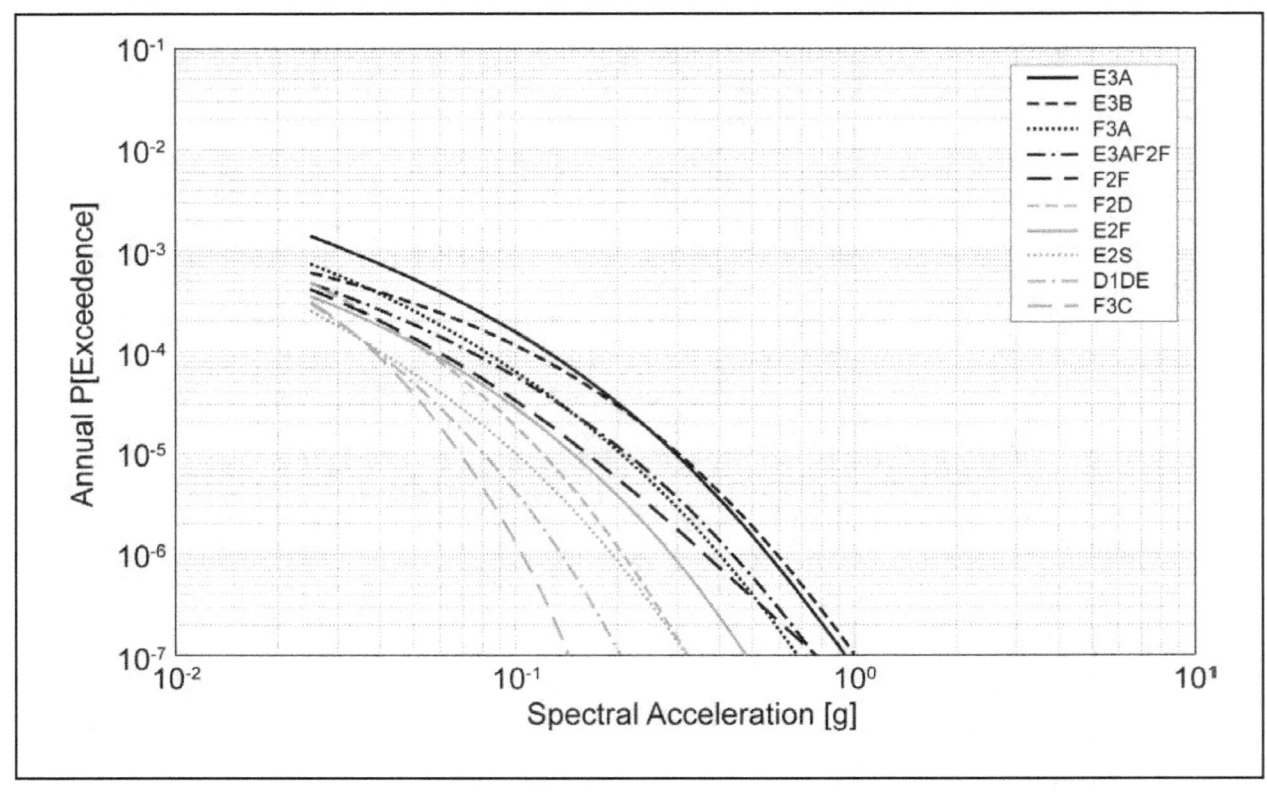

Figure 4-9. Example of Sensitivity Analysis Showing the Contribution of Various Seismic Sources to the Seismic Hazard at a Site.

4.9 Final Hazard Calculations

Based on consideration of the data and feedback provided throughout the project, the EEs will develop their final models, the models will be documented in the final HID, and final hazard calculations will be conducted. To ensure that the final model is "locked down" and that no changes will be made during the report preparation phase of the project, it is suggested that the development and delivery of the HID by the evaluator to the hazard analyst mark the official end of the model finalization effort. The products from the final hazard calculations should include all of the visualizations (e.g., plots and figures) and outputs that are needed for the subsequent use. In addition to the hazard outputs, a series of sensitivity analyses should be conducted—similar to those conducted for purposes of feedback—to provide a basis for understanding the dominant contributors to the hazard results and to the associated uncertainties. Section 5.1 provides a discussion of the recommended products resulting from hazard calculations.

4.10 Documentation

The original SSHAC report devotes a chapter (Chapter 7) to recommendations regarding documentation, and those recommendations are endorsed here. The project documentation is the fundamental basis for the reader to understand (1) what process was used in the hazard analysis; (2) what data were available and used in the evaluation process; (3) how the data, models, and methods of the larger technical community were considered; (4) the elements of the models and their technical bases; (5) how the models capture the center, body, and range of technically defensible interpretations; and (6) the hazard results and instructions for their use. The draft report is subject to review by the sponsors and other groups, but especially by the PPRP.

Although the focus of NUREG/CR-6372 and this document is Level 3 and 4 assessments, Level 2 updates to Level 3 and 4 regional models are commonly used for nuclear regulatory actions. These Level 2 assessments, particularly those used in regulatory submissions, should follow the below guidance and should produce, clear, complete, and transparent documentation such that the updated model can be shown to capture the CBR of the TDI, which is the objective of all SSHAC based assessments.

4.10.1 Process Used

As discussed in Section 2.2, a primary finding of the SSHAC study was that the process followed in a hazard analysis can have a significant impact on the hazard results. Accordingly, both the original SSHAC report and this NUREG are intended to provide guidance on elements of a process that can help provide stable hazard results if conducted in the manner recommended. It is therefore important that the project report provide a detailed description and explanation for the process conducted. An explanation should be given for the selection of the SSHAC Level and why it was deemed appropriate for the study. The discussion should include the activities, workshops, participants, schedules, and organizational structure used to achieve the project deliverables. A description and basis should be given for any revisions or refinements to the SSHAC process outlined in this NUREG.

For Level 4 studies, a description should be given for the approach used to combine or integrate the individual evaluator expert assessments including a justification for why that approach is appropriate for the particular project circumstances. For example, previous Level 4 studies have usually combined the EE assessments using equal weights. This approach has been justified on the basis of having provided all experts with equal access to all databases, equal exposure to all proponent models and alternative viewpoints, clear guidance throughout the project that all experts are required to fulfill the role of evaluators, etc. If an alternative method is used for combining the expert assessments (e.g., differential weighting), the details and justification for the approach should be documented.

4.10.2 Data Considered

It is important for the reader of a project report, especially one reading the report some years after the project was completed, to understand fully what data were considered at the time of the study and how those data were used by the TI or EE experts in their evaluations. This includes the data that were made part of the project database and new data collected as part of the study. A summary of the project database should be included or appended to the project report, or reference should be made to a separate document. The database description should include a description of the metadata to allow for an understanding of the specifics. In some cases, the project database will be considered a deliverable of the project and will be expected to be in a format for subsequent use and perhaps continued update. Release of the database will be project specific and should be specified in the project contract terms.

It is important to document and inventory all data that were considered in the course of the project including those data that were not used. For those data that were relied upon, it is important to also document the manner in which those data were used. A useful example of this type of documentation was developed as part of the CEUS SSC project. Two types of tables were developed: Data Summary tables (an example of which is provided as Table 4-5) and Data Evaluation tables (an example of which is shown in Table 4-6). The Data Summary tables provide a documentation of all data that were considered by the TI team members in the evaluation. In addition to the citation, the Data Summary tables provide a summary of the potential relevance of the reference to the task at hand (e.g., SSC or GMC). The Data Evaluation tables document those data that were relied upon in the expert assessments. The tables summarize assessments made by the TI or EE members regarding the quality of the data, the specific issue (e.g., seismic source) that was addressed with the data, and the degree to which the data was relied on in the assessments. These tables, of course, must be supplemented with a detailed description in the text of the models and assessments made, but they are an efficient and effective way to summarize the databases for all of the assessments.

The tables shown are one approach to documentation of available data, models and methods that is particularly useful for describing data relevant to SSC models. However, other approaches may be used depending on project needs. For example, GMC projects, which use data, models and methods very differently, may find it more appropriate to use other forms of documentation to meet the above stated goals. In particular, the documentation should describe the full suite of information available at the time of the study and should also detail the complete set of information that is used in developing the model along with its use and the degree to which it was relied upon.

4.10.3 Elements of the Models

The technical discussion of the elements of the expert models and their technical bases is the backbone of the report. Each element of the models (i.e., the logic-tree branches and weights) should be discussed in detail including all models, parameters, and uncertainties. The discussion should include all data, models, and methods that were considered including those that may have been evaluated to not be credible enough to be included in the final model (i.e., those that are given zero weight). A key part of the model documentation process is the assessment of how the final model is believed to capture the center, body, and range of the technically defensible interpretations.

The documentation process will differ somewhat for Level 3 and 4 studies. The discussion of the model for a Level 3 study will be final model developed by the TI team. In a Level 4 study, each EE should prepare a report that summarizes all elements of their individual models along with the technical basis for all model components. It is recommended that the TFI provide the EE panel an outline to use in documentation to increase the likelihood that all necessary topics are covered and that the EE reports are consistent in format. These individual EE reports should be appended to the final project report. The TFI should prepare a summary in the main body of the report of the assessments made by all of the evaluator experts, drawing conclusions regarding common elements across the EE assessments or differences, as appropriate.

4.10.4 Hazard Results and Instructions for Their Use

The project report should include a thorough documentation of the hazard results, sensitivity analyses that provide information on the dominant contributors to the hazard, and instructions for the use of the hazard results for the anticipated users. The latter may include any caveats or limitations to the use of the results. For example, some hazard results may be limited to certain annual frequencies of interest or geographical limits to their applicability.

Table 4-5. Example Data Summary Table from the CEUS SSC Project

Citation	Title	Relevance to SSC
GEOLOGIC STRUCTURES INTERPRETED FROM GEOLOGIC, GRAVITY, MAGNETIC, AND SEISMIC-PROFILE DATA		
Mooney et al. (1983)	Crustal Structure of the Northern Mississippi Embayment and a Comparison to Other Continental Rifts	Information on the deep structure of the northern Mississippi embayment, gained through an extensive seismic refraction survey, supports a rifting hypothesis. The confirmation and delineation of a 7.3 km/s layer, identified in previous studies, implies that the lower crust has been altered by injection of mantle material. The results indicate that this layer reaches a maximum thickness in the north-central embayment and thins gradually to the southeast and northwest, and more rapidly to the southwest along the axis of the graben. The apparent doming of the 7.3 km/s layer in the north-central embayment suggests that rifting may be the result of a triple junction located in the Reelfoot basin area.
McKeown et al. (1990)	Diapiric Origin of the Blytheville and Pascola Arches in the Reelfoot Rift, East-Central U.S.: Relation to New Madrid Seismicity	Earthquakes in the NMSZ correlate spatially with the Blytheville arch and part of the Pascola arch, which are interpreted to be the same structure. Both arches were formed by diapirism. The rocks in the arch are more highly deformed, and are therefore weaker, than adjacent rocks. Seismicity is hypothesized to be localized in these weaker rocks.
Nelson and Zhang (1991)	A COCORP Deep Reflection Profile Across the Buried Reelfoot Rift, South-Central U.S.	Deep reflection profile line reveals features of the late Precambrian (?)/early Paleozoic Reelfoot rift. The Blytheville arch, an axial antiformal feature, as well as lesser structures indicative of multiple episodes of fault reactivation, is evident on profile.
Hildenbrand and Hendricks (1995)	Geophysical Setting of the Reelfoot Rift and Relations Between Rift Structures and the NMSZ	Provides discussion of several potential field features inferred from magnetic and gravity data that may focus earthquake activity in the northern Mississippi embayment and surrounding region. Summarizes complex tectonic and magmatic history of the rift.
Potter et al. (1995)	Structure of the Reelfoot–Rough Creek Rift System, Fluorspar Area Fault Complex, and Hicks Dome, Southern Illinois and Western Kentucky—New Constraints from Regional Seismic Reflection Data	Interpretation of an 83 km segment of seismic reflection data across the northern part of the Reelfoot rift—Fluorspar Area fault complex (FAFC)—in southeastern Illinois and western Kentucky. Notes that NMSZ appears bounded on the north and south by Cambrian accommodation zones that linked segments with differing rift geometry. A series of grabens and horst in the FAFC document a late Paleozoic reactivation of the Cambrian rift. Beneath two of the FAFC grabens, the bounding faults meet within the Knox Group and do not continue to depth. Other normal faults in the FAFC clearly offset the top of Precambrian basement.

This table is used to identify all data sources considered for a particular project. Note that in addition to the reference citation, a summary is given of the relevancy of the document to the assessment being made, in this example for seismic source characterization.

Table 4-6. Example Data Evaluation Table from the CEUS SSC Project

Data/References	Quality 1=low, 5=high	Notes on Quality or Data	Source considered e.g., A, B	Used in SSC and reliance level 0=no, 5=high	Discussion of data use	In GIS database
Instrumental Seismicity						
CEUS SSC database	5			1	Reviewed for alignment of microseismicity	Y
Chiu et al. (1997)	3	Peer-reviewed journal article	EMF_S	2	Analysis of seismicity suggests there is an active fault source along the southeastern flank of the Reelfoot rift.	N (seismicity discussed in paper is not specifically included as a GIS layer, but M≥3 earthquakes along the margin are included in the CEUS SSC earthquake catalog)
Historical Seismicity						
Hough and Martin (2002)	3	Analysis of sparse intensity data for a large aftershock of the 1811 earthquake	EMF_S	2	Aftershock of December 1811 NMSZ earthquake (NM1-B) - M 6.1 ± 0.2, location of event not well constrained, but probably beyond the southern end of the NMSZ, near Memphis, Tennessee (within the southwestern one-third to one-half band of seismicity identified by Chiu et al. (1997)).	N

This table is used to describe the manner in which data have been used in the hazard analysis. The table includes an evaluation of the quality of the data and the degree of reliance made in the analysis.

4.10.5 Document Review

Once the draft project report has been completed, it is subject to review by the PPRP and, if applicable, the sponsors and other groups. The PPRP's review is vital to the SSHAC process and, because the panel will have undertaken a participatory review throughout the project, they will have a working familiarity with all aspects of the project, and they will be in a strong position to comment on the degree to which the documentation has sufficiently and completely represented the project. It is suggested that the PPRP be provided by the Project Manager with a set of review criteria against which they should review the report. Just as their review throughout the course of the project, the PPRP review should entail both technical and process aspects.

Example review criteria are the following:

Technical

- Have all data used in the assessment been identified and documented?

- Have all elements of the model been defined in sufficient detail?

- Have the model elements and expressions of uncertainty (e.g., logic-tree branches and their weights) been technically justified?

- Are there any technical issues that have not been sufficiently addressed?

Process

- Has the choice of SSHAC Level been justified?

- Have all of the essential steps of a SSHAC process been followed and documented?

- Is the data evaluation process sufficiently justified?

- Is there clear documented evidence that the views of the larger technical community have been considered?

- Has the integration process been sufficiently documented such that the center, body, and range of technically defensible interpretations are well justified?

Other reviewers, such as the sponsors, may have other review criteria such as provision of the project deliverables, clarity in descriptions of their use, and the degree to which users of the report will understand its contents.

4.10.6 Peer Review Documentation

As discussed in Sections 3.6.8, 5.5, and elsewhere, the participatory peer review process is an important component of SSHAC Level 3 and 4 projects. The selection of the PPRP should occur early in the process (Section 4.4), and the PPRP should provide its review and feedback periodically following key points in the projects. This should include written comments following all workshops and other decision points indicated by the PTI and Project Manager. In some cases, the PPRP have also reviewed the Project Plan.

Because the feedback typically includes different types of comments with different levels of urgency or expected response, using a set explanative notation to classify the comments provided has been found to improve communication and highlight what the PPRP believes to be critical issues. In the case of the workshops, it has also been found to be useful to allow for an informal daily debriefing to allow for mid-workshop corrections and to allow for discussion of comments expected to appear in the workshop report. In some cases, the discussions held in the debriefing are sufficient to completely close out a comment or concern. However, often it is also appropriate to included comments in the PPRP's workshop report for completeness and transparency.

It is also useful for the TI Leads and the PTI to provide their responses to the PPRP comments in writing to ensure that all parties are aware of the issues and the manner in which the project will respond. The Draft Project Report should be a review product for the PPRP, and comments should be provided in writing to the project team. At the conclusion of the project and after finalization of the final report, the PPRP should provide its final comments. These final comments should include the PPRP's final evaluation of whether the TI/EE teams have considered the technical community's viewpoints and have made a concerted attempt to capture the center, body, and range of technically defensible interpretations in their final models. The comments should also address their final assessment of the process followed by the project and whether or not that process is acceptable as a SSHAC process. The PPRP's assessment should be included in the final report of the project.

5. PRACTICAL IMPLEMENTATION OF PROCESSES

The previous two chapters have outlined the essential components for conducting a Senior Seismic Hazard Analysis Committee (SSHAC) Level 3 or Level 4 Probabilistic Seismic Hazard Analysis (PSHA). This chapter discusses additional considerations regarding practical implementation of such a process to ensure a successful outcome, including highlighting some pitfalls that can be easily avoided when conducting the study but which can be difficult to fix at a late stage. The chapter also makes a number of recommendations that go beyond the original SSHAC Guidelines and reflect some of the lessons learned from the experience of application. As always, a successful outcome here implies a complete and well-documented hazard study that contributes to regulatory assurance that the hazard has been robustly evaluated and that the associated uncertainty has been adequately captured.

5.1 Define Required Goals and Outputs of PSHA

A PSHA is always conducted for a specific purpose, ultimately linked to mitigation of earthquake risk to engineered structures or facilities. From the very outset of the project, at the planning phase, it is therefore strongly recommended to engage with the sponsor to define the required deliverables. This will usually necessitate dialogue not only with managers from the sponsoring organizations but also with those who will make use of the output from the hazard calculations.

The SSHAC guidelines provide clear specification of requirements for how the hazard output should be presented and documented, and Section 4.10 provides additional discussion. The basic outputs required from a PSHA study are clearly specified in Regulatory Guide RG 1.208 (USNRC, 2007). As a minimum, a site-specific PSHA should provide the following representations of the ground-shaking hazard:

- Hazard curves for each of the required ground-motion parameters including the mean and several fractiles (5, 15, 50, 85 and 95%) over a wide range of annual frequencies of exceedance.

- Curves showing the contribution of individual seismic sources to the mean hazard for a range of response frequencies.

- Uniform hazard spectra at specified annual exceedance frequencies, which will generally include 10^{-4} and 10^{-5}.

- Disaggregation of the mean hazard at selected response frequencies and annual exceedance frequencies to identify the contributions from different bins of M, R and ε (see Appendix Section B.3.3).

- Results of sensitivity analyses showing contributions to the variance in the hazard curves (i.e., epistemic uncertainty) from different sources of uncertainty.

Other representations of the hazard and design ground motions that may be required—such as conditional mean spectra (Baker and Cornell, 2006) and acceleration time-histories—can all be obtained from post-processing of the above information. A very important issue to define *a priori* is the lowest annual frequency of exceedance for which ground motions may be required. With modern computing capacities, it is a relatively straightforward task to extend the calculations to lower exceedance frequencies whereas considerable logistical difficulties can arise if the need for these only comes to light after the project is completed. Extrapolation of calculated hazard curves is not an acceptable approach, so the lower limit should be carefully considered before the hazard calculations commence. Whereas for design of nuclear power plants the annual frequencies of exceedance of interest will generally be in the range from 10^{-4}

to 10^{-5} (e.g., USNRC, 2007), for a complete probabilistic risk analysis (PRA), hazard estimates may be required as low as 10^{-8}. Extending PSHA to such low annual exceedance frequencies is much more involved than a simple extrapolation of routine hazard models because it will influence the models considered. It is also important to avoid misunderstanding that such PSHA is not an extrapolation to some distant time in the future but rather to feasible events with very low likelihoods within the present tectonic setting[13].

Another issue that should be identified at an early stage is the target horizon at which the ground shaking hazard is to be computed, often defined by an elevation relative to an established datum or a corresponding average shear-wave velocity over the 30 meters directly below that level, $V_{s,30}$ for a site-specific study. This may depend on the specific technology and the associated embedment of the foundations as well as on the analyses that are to be performed, but in general it will be a stiff layer at some depth below the original ground surface at the site. The selection of the reference geological stratum can also be an interface issue, as discussed in Section 5.9.

From an engineering perspective, the issue is to define the ground-motion parameters that will be needed for the analyses to be conducted. This will generally be determined by the way that the ground shaking is represented in the fragility functions for structures, systems, and components (SSCs) of the plant or the component of motion considered for design. At a minimum, the ground shaking will always need to be represented in the form of a response spectrum of pseudo-absolute acceleration. The range of associated response frequencies that need to be covered by the response spectra must be clearly identified. The SSHAC guidelines list a small number of response frequencies for which the hazard should be calculated, but with modern computing capacities, there is no need to limit the output only to these target response frequencies. An issue that may need to be addressed is whether coefficients or spectral ordinates will need to be interpolated for any ground motion prediction equation (GMPE) that does not provide predictions at any of the specified response frequencies. Such interpolation is generally straightforward, but difficulties may arise if response frequencies are specified beyond the limits appropriate for each GMPE as a result of record processing issues (e.g., Akkar and Bommer 2006; Akkar et al., 2011),

For the horizontal component of motion, it needs to be clear how the two orthogonal components from each accelerogram are treated in the fragility analyses and for the output to be expressed according to the same convention. The most commonly used horizontal component definition in current ground-motion prediction equations is the geometric mean of the two horizontal components for each record or some variation of this definition. Empirical relationships can be used to transform spectral accelerations defined in this way to other component definitions (e.g., Beyer and Bommer, 2006). There are two options for derivation when the response spectrum of the vertical component of motion is also required. One is to conduct a parallel PSHA using equations for the prediction of the vertical component motion, which was the approach recommended in the original SSHAC guidelines: "*If important, we recommend that vertical motions be obtained from independent analyses, in the same manner as for the horizontal motions*" (NUREG/CR-6372). The second option is to obtain the vertical spectrum through the application of appropriate V/H ratios to the horizontal spectrum (e.g., McGuire et al., 2001; Bozorgnia and Campbell, 2004a). The second option is now considered preferable, because the former approach can result in horizontal and vertical components of motion controlled by very different earthquake scenarios.

---

[13] This confusion is sometimes increased by referring to such low annual frequencies as return periods; for example, as the "1 million year ground motion".

Another engineering requirement that is best discussed at an early stage is the damping ratios that may be required for the structural analyses. Predictive equations are invariably derived for spectral ordinates with 5 percent of critical damping. However, for many engineering applications, other values—both higher and lower than this nominal level—may be relevant. Response spectra for other damping ratios can be obtained by applying scaling factors to the 5-percent-damped acceleration ordinates, but these factors should be selected to reflect the influence of either magnitude and distance (Cameron and Green, 2007) or of duration (Stafford et al., 2008a).

If any of the engineering applications require other parameters, such as peak ground velocity (PGV) or duration, this should be made clear up front so that a consistent approach can be adopted for their determination.

5.2 Organizational Structure and Management

The original SSHAC guidelines give only sparse guidance on practical issues of structural organization and management of a major SSHAC project, but they are of clear importance to ensuring a successful outcome. The recommendations made here are based on experiences of what has and has not worked well in practice. These recommendations may need to be adapted to the requirements and the context of each project.

Central to the successful execution of a SSHAC Level 3 or 4 project is clarity regarding the roles and responsibilities of each participant as well as their required attributes, which determine their conduct in the project. For this reason, it is recommended that the specific role of each participant be clearly identified in their contract, together with a list of the expected attributes and assigned responsibilities. Some of the basic considerations that underlie these proposals for the organizational structure of a Level 3 or Level 4 PSHA project are as follows:

- The technical teams require a degree of autonomy provided that it can be demonstrated that the project is being executed to plan, including conforming to schedule and budget.

- At the same time, the project sponsors have a right to continuous oversight of the project and need a clear channel through which to communicate with the teams executing the project. A dedicated, full-time Project Manager (PM) is needed who is the point of contact between the project and the project sponsor. The PM is responsible for ensuring adherence to scope, schedule and budget. The PM develops contracts with all technical personnel and subcontractors, organizes the workshops (including issuing invitations to all participants and observers), and keeps the sponsors apprised of progress in terms of scope, schedule, and budget. The responsibilities of the PM include holding each participant to their contractual roles and responsibilities.

- The technical leads on the project are responsible for all technical aspects of the project and, therefore, should not be involved in the administrative duties described for the PM in the previous bullet point but at the same time should be able to call on the support of the PM for administrative issues as required.

- On all complex technical issues that arise, from the sponsor or from the PPRP, the PM should defer to the technical leader of the project (the PTI or PTFI, roles that are described below).

- At the top level of the project, technical leadership is required from an individual with broad expertise and experience in PSHA, working alongside the project manager but with a clear division of labor between administration and technical orientation and decision making (see discussion below of PTI and PTFI roles).

- This same individual needs to be part of the communication between the sponsor and the PM regarding questions and concerns of a technical nature.

In light of these considerations, the first recommendation made regarding project structure is that a position should be created that is designated as Project TI (PTI) Lead in a Level 3 study and as Project TFI (PTFI) in a Level 4 study, as noted in Chapter 3. This position is the principal point of contact with the technical aspect of the project. As such, this individual must possess a broad and deep knowledge of the practicalities of conducting a PSHA, and be prepared to commit a great deal of time and energy to the project. There is no reason why the PTI or PTFI cannot be the same person as either the SSC TI/TFI Lead or GMC TI/TFI Lead, provided they possess the required expertise and can commit the required time to the project.

The PTI or PTFI, the PM, and the representatives of the sponsors should stay in close communication to provide a forum for reporting on progress, exchanging information, raising queries, and expressing concerns. However, the clear legal ownership of the project and ultimate executive authority must reside with the project sponsor. Figures 5-1 and 5-2 illustrate how a Level 3 and Level 4 project, respectively, may be organized. In both cases, the scheme is for a full PSHA and clearly these would need to be adapted for projects that are focused only on either seismic source characterization (SSC) or ground motion characterization (GMC) issues.

The roles of the PTI or PTFI and the PM are complementary. However, the overall technical direction of the project, in terms of the development of the logic-trees and the execution of the PSHA calculations, should come from the PTI or PTFI. The PTI or PTFI will work closely with the TI Leads or TFIs and will be supported in all administrative and logistical aspects by the PM. This is a refinement with respect to the original SSHAC guidelines that identified, somewhat vaguely, a 'Project Leader' defined as follows: "*The Project Leader (often one individual, but possibly a small team) is the entity that takes managerial responsibility for organizing and executing the project, oversees all other project participants, and 'owns' the study's results in the sense of assuming intellectual responsibility for the project's overall technical validity*" (NUREG/CR-6372). Elsewhere, the guidelines refer to "*the Project Leader and the TI (or TFI)*" suggesting that they were viewed as separate entities. Herein it is recommended that the Project Leader role should actually be that of an overall TI Lead or TFI, with a scope along the lines defined in the above quote from the SSHAC guidelines but with the managerial responsibilities belonging to the PM.

In a Level 3 study (Figure 5-1) careful thought must be given to the composition of the TI teams for the SSC and GMC subprojects. If possible, the teams should not be dominated by personnel from a single institution or company because there could be a perception that this will not provide sufficient diversity of viewpoints and approaches.

In some Level 3 projects, a distinction has been made between TI team and TI staff, with the implication that those in the second group contribute to data collection, processing, and analyses but not in the decisionmaking as evaluators. The concept of TI staff is not useful because these individuals fully participate in the technical integration process that builds the logic-trees. It is, therefore, recommended that this term should not be used. Membership of the TI team, in which each and every member is expected to act as an evaluator expert and share ownership of the technical decisions, should be identified from the outset of the project. Other participants who are not members of the TI team should be designated as specialty contractors.

A point to be borne in mind while populating project teams is that a need exists to look to the future and to use SSHAC Level 3 and 4 projects to prepare future project leaders. Although some of the technical expertise required can be obtained in Level 1 and 2 studies as well as in research, there is no better way to learn how Level 3 and 4 studies are conducted than through participation in real projects. Therefore, the demographic distribution of project teams should ideally include both those bringing many years of experience and those who may be future TI leads and PPRP members.

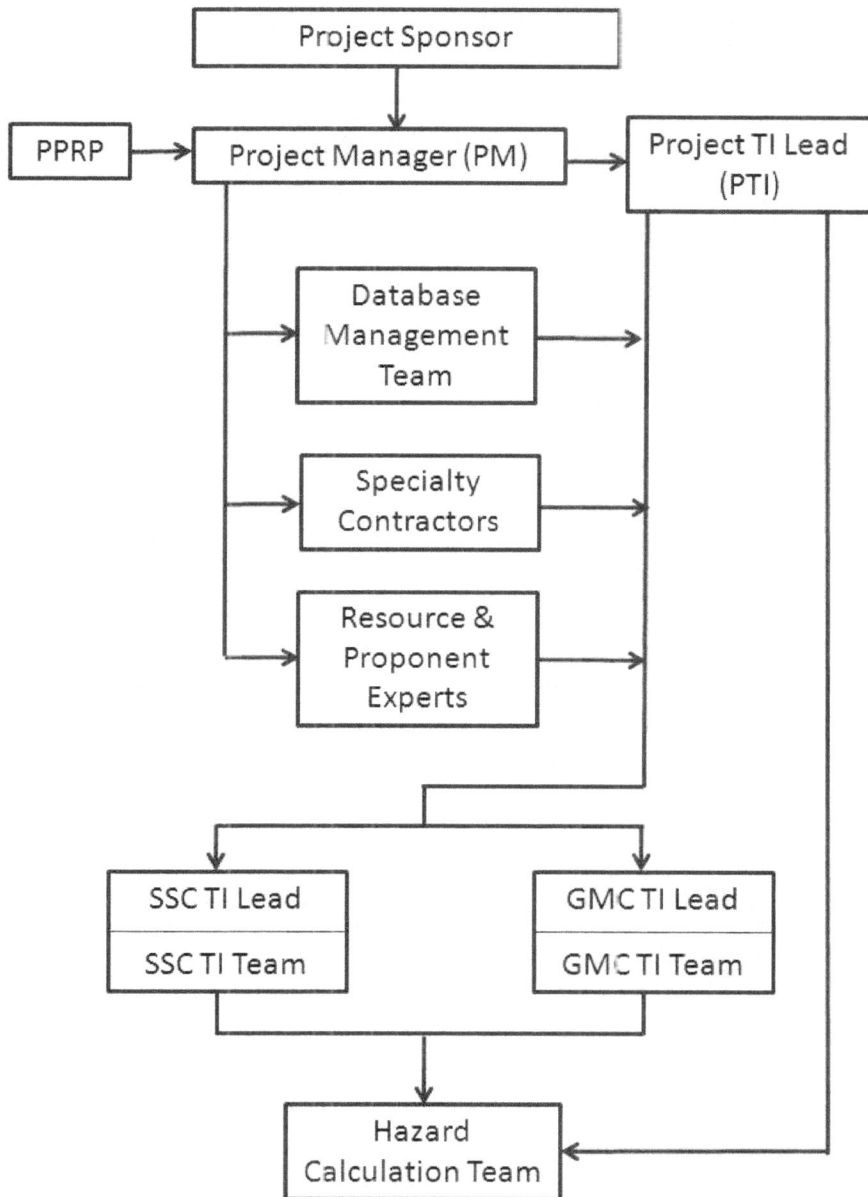

Figure 5-1. Project Organization Structure for a Level 3 PSHA.

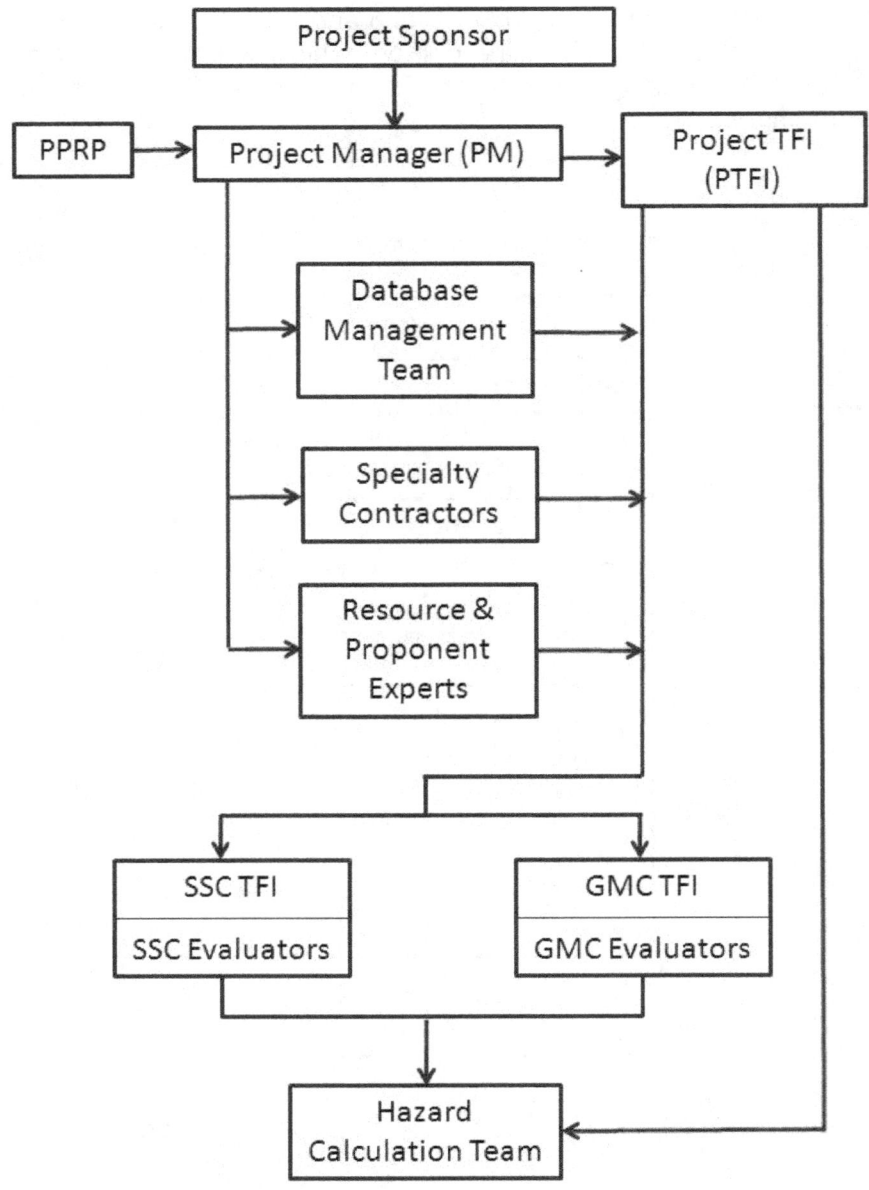

Figure 5-2. Project Organization Structure for a Level 4 PSHA.

5.3   Role of Regulators and Sponsors

The discussions of the project organization and management in the previous section highlighted the importance of engagement with and by the project sponsors. In each project, different structures may be established depending on the number of sponsoring agencies and the desired forums for interaction. If there are several sponsors, it is particularly important that a structure is established for coherent and coordinated communication with the project. Regardless of the specific structure adopted, the sponsors should have a single point of contact

94

with the project, namely the PM. The PM must ensure that technical direction comes from the PTI or PTFI. All technical participants (e.g., evaluators, specialty contractors, resource and proponent experts) therefore need to communicate all technical concerns to this specific individual. It may be necessary at certain times for the PM to facilitate direct communication between the sponsor and the PTI/PTFI, but the PM should be party to such communication. The sponsors' representatives should engage with the project from the outset, particularly in establishing deliverables and schedule. The sponsors' representatives should also attend all workshops as observers as well as receiving copies of PPRP reports and the responses from the TI Leads or TFIs. At the same time, it is also reasonable to expect that, if the sponsor has approved the project plan including budget, schedule, and the key appointments (TI Leads or TFI, PPRP members, etc.), the sponsor will delegate all responsibility for the day-to-day management of the project to the PM including responsibility to maintain budget and schedule and regular reporting of progress of the project to the sponsor. It is not appropriate for the sponsor to attempt to direct the technical elements or outcomes of the project.

A common use of SSHAC Level 3 and 4 hazard studies is to contribute to licensing or safety evaluations of safety-critical facilities. There is therefore great advantage to be gained if the NRC or other relevant regulatory body follows the entire process, primarily by attending the workshops as observers. Although observers in a SSHAC workshop are precluded from the technical discussions, it is suggested that a specified time be allotted at the end of each day or each workshop to open the floor to questions and comments from observers. In such a context, the regulator could provide feedback, raise concerns, or ask for points of clarification. In addition, the regulator can be provided with a copy of the project plan for information. It is suggested that the choice of the SSHAC study level (see Sections 3.3 and 4.2, and Chapter 6) will be made in consultation with the regulator.

The most beneficial outcome of such engagement between the project and the regulator would be enhanced assurance associated with the assessment of seismic hazard at the site, which can lead to reduced times associated with regulatory reviews. Key to such an outcome is the confidence and reassurance that may be provided by the PPRP who, as impartial and experienced recognized experts in the field of PSHA, will be able to give the regulator confirmation that a proper process was followed and that sound technical assessments have been made.

5.4   Conduct of the Workshops

The formal workshops (minimum of three) are key events in a SSHAC Level 3 or 4 study (as discussed in Chapter 4) although it is important to always keep in mind that the actual work of conducting the seismic hazard analysis is done between (rather than in) the workshops. The workshops are opportunities for outside participants to inform the expert evaluators, for reporting the progress to date, and for identifying the issues still to be addressed. The workshops are also important because they allow for the process of assessing the range of technically defensible interpretations to be conducted formally and under observation. The workshops will invariably be relatively short in duration, at most a few days and typically not longer than 5 days.

It is therefore important that the best possible use be made of the available time. For this reason, clearly briefing the speakers can be useful. This can include coaching of the speakers on appropriate conduct. The objective of this is to ensure that the presentations are clearly focused on the topic assigned and to what is actually relevant to the seismic hazard assessment. All of the presentations should be appropriate to the particular workshop, which specifically means at Workshop 1 (Hazard Significant Issues and Available Data) that speakers should avoid interpretation of data or models in terms of their hazard implications. Conversely, at Workshop 2 (Discussion of Alternative Models) speakers should focus specifically on interpretations of data and models in terms of issues that are important to the seismic hazard. In some cases, it has been useful for the TI/TFI to provide the speakers with a list of questions that the speaker is expected to address. These questions focus the speaker on aspects of the data, model, or method that the evaluator experts need to understand and to help focus subsequent discussion.

The workshops also provide an appropriate opportunity for the technical leaders[14] to remind evaluators of the importance of being aware of cognitive biases and the need to avoid allowing these biases to influence their assessments (see Section 5.11). These issues may be raised at Workshop 1 so that the participants are aware from the start of the project; however, the time when they most need to be reinforced is Workshop 2 prior to the model-building process by the evaluator experts.

In many projects, particular those with international participation, the workshops will provide the primary opportunity for the PPRP to interact with the project participants. This leads to particular consideration of how to ensure that the PPRP engages most effectively at the workshops. All members of the PPRP should be present at all project workshops, but for practical reasons this may not always be possible. For this reason, a quorum may be established for the PPRP on each project but with the proviso that at each SSC or GMC workshop at least some of panel members who are experts in the respective areas should be present.

Although during the formal workshop proceedings members of the PPRP act as observers and do not participate in discussions, it is important that they are given adequate opportunities to raise issues and pose questions to the evaluators.

These opportunities can be created in a number of ways including the following:

- Opening the floor to observers to make comments at end of each day of the workshop.

- Holding informal debriefing sessions between the PPRP and key project participants (PM, TI Leads and PTI or TFIs and PTFI) after the close of each day of the workshop.

- Extended debriefing meeting between the PPRP and key project participants after the close of the workshop.

These are all useful ways to allow the PPRP to raise its concerns in a timely fashion. In many cases, discussion at the end of each session or in a daily debriefing have allowed issues to be addressed and resolved during the course of the workshop. The written reports from the PPRP after each workshop also provide the expert evaluators with information to be taken into account as the work progresses. Therefore, it is appropriate for completeness of the documentation if some of the comments made in the debriefings are included in the PPRP report as well. The TI Leads should respond in writing to the PPRP comments to document that the comments are understood and the manner in which the advice given will be addressed.

Workshop #3 is important in that it provides the first opportunity for all participants including the PPRP to see the full preliminary hazard model (see Section 4.6.3)[15]. After the workshop, the PPRP's next task is reviewing the draft final of the PSHA report. It is problematic if serious technical challenges arise during the review of that document. However, there is greater risk of

---

[14] These leader are the PTI or TI lead in a Level 3 assessment and TFI or PTFI in a Level 4 assessment.

[15] This discussion assumes that only the three workshops necessary to meet the minimum requirements are held.

this occurring if the PPRP does not have ample opportunity to raise questions and concerns at Workshop #3. For this reason, it is recommended that at Workshop #3 the PPRP be relieved of their observer status and allowed to participate directly in discussions and technical challenge of the preliminary models developed by the evaluation and integration teams. At Workshop #3, the main participants will be the evaluators/integrators, and the PPRP. A possible structure for the workshop could then be that the mornings are used to present hazard sensitivity feedback for discussion by hazard evaluators/integrators and in the afternoons the PPRP be given the chance to interrogate the models and their technical bases. In this way, all issues can be raised and thoroughly discussed before the TI teams finalize the model and draft the PSHA report. It is beneficial if the PPRP's questions and concerns are kept in mind while documenting the study. This should contribute to fewer questions and concerns being raised by the PPRP in their review of the draft final report.

A final point regarding the workshops concerns the independence and impartiality of the PPRP, and particularly the perception of this objectivity by observers. Given the relatively small and close-knit nature of the specialist technical communities in the earth sciences, it is very likely that members of the PPRP and the technical staff on the project will know each other and may even be working together on other endeavors. Therefore, there should be no unrealistic expectations about separation and distance, but all members of the panel should be vigilant about being drawn into participating in the actual technical assessments. Depending on the situation and environment in which the study is connected, project organizers may consider taking some measures to avoid excessive social and informal interaction between the members of the PPRP and other project participants. Options may include accommodating the PPRP members in a separate hotel to simply arranging a separate table for them to take meals during the workshop. However, this separation does risk limiting the PPRP's ability to observe discussions and interactions taking place at the entire workshop. It may be entirely appropriate to also facilitate interactions at these times between the PPRP and either the sponsor or the regulator.

5.5 Consensus Among the PPRP

The PPRP is required to submit written reports after each workshop and after issue of the final PSHA report; it is highly desirable that these be consensus documents reflecting the views of the panel as a whole. Every effort should be made to avoid arriving at the end of the project with a "minority view" among the PPRP members. This makes the role of chair of the panel very important. As a result, this position needs to be held by someone who is well regarded by their colleagues and who possesses the personal qualities to encourage constructive interaction, facilitate effective dialogue, and resolve differences of opinion. The PPRP Chairperson should be someone with extensive experience with PSHA projects and in participating in review panels.

Experience has also shown that the work of the PPRP in drafting these consensus documents is much more easily and effectively carried out in person rather than through remote correspondence. Therefore it is recommended to make provisions for the PPRP to meet immediately before each workshop and again immediately after the workshop concludes. The purpose of the first meeting is to discuss the project and any concerns and to identify particular issues that the panel will wish to focus on during the workshop and during the debriefing sessions during the workshop. Prior to Workshop #1, the panel may want to review the format of other PPRP's reports and consider an approach for documenting their own reviews. This facilitates the compilation of PPRP comments throughout the workshop. The meeting at the end of the workshop is expressly assigned to the drafting of the consensus report of the panel giving feedback on the workshop and the progress up to that point. There are great benefits in the PPRP report being drafted before the panel members depart, and a PPRP may establish a rule that *de facto* acquiescence on the report will be assumed on the part of any panel member who is absent or who leaves before the report is completed.

There is an onus on the PPRP to identify and communicate any potential problems, whether technical or procedural, as early and as energetically as possible. It is important that the panel members and, in particular, their chairperson feel free to express their concerns directly and in forthright fashion. The panel should also expect to receive written responses from the PM, TI Leads, TFI, PTI or PTFI, as appropriate, when these concerns are expressed in writing.

The proposal to positively encourage the PPRP to raise questions and pose technical challenges to the evaluators/integrators in Workshop #3 (see Section 5.4) should serve to help avoid the possibility of minority views arising. However, if a minority view does emerge within the PPRP at the end of the project, then it is important that it be clearly documented so that its significance can be weighed by both the sponsor and the regulator. The first thing that should be clarified is whether the minority position is with respect to the process review or the technical review (or both). This is important because if the PPRP reaches consensus that the project did adhere to the requirements of a SSHAC Level 3 or 4 process (even if there is dissention regarding technical issues), then issues related to the conduct of the study can be laid to one side and the technical issues addressed separately. The PPRP should also convey—without necessarily revealing individual identities—how many of the panel members share the minority view.

5.6 Evaluation of Existing Models and Building New Models

As explained in Sections 3.1 and 4.2, the SSHAC process is divided into two stages: first evaluation (of existing data, methods and models) and then integration of models into SSC and GMC logic-trees that form the basis of the PSHA calculations. It can be stated with some confidence that very rarely will the evaluation phase identify a set of existing models that can provide all the input to the PSHA, requiring only that the experts select from among these models to populate the logic-tree branches and assign weights to these branches in the integration phase. For SSC projects, it is virtually never the case that, in the technical literature, sufficient and adequate seismic source models are available for the specific location under study, although preexisting models of various source components may exist. So, in practice, any SSC project will build new models describing the location of seismic sources and their recurrence characteristics. For GMC projects, one might expect that the process may only involve selection of published GMPEs, but in practice this is unlikely to be the case. With the exception of a few geographical regions for which there are several GMPEs already published, GMC logic-tree development will require bringing in models from other host regions and applying adjustments either to render the models more applicable to for local conditions or simply to make the models compatible in terms of predictor variables (see Section 5.9, for example). Therefore, even if the evaluators begin with the intention of only using published prediction equations, these are likely to undergo transformations that—in effect—render them into new models. For low-seismicity regions, models will often be brought in from other regions due to the lack of an indigenous strong-motion database. If stochastic GMPEs are adopted from another stable region, such as Eastern North America, then consideration should be given to the benefit and drawbacks of adopting models extrapolated from weak-motion recordings in another region as compared to developing new models from local weak-motion recordings, if these are available.

In short, if the evaluation phase is likely to conclude that existing SSC and GMC models are insufficient to fully define the input to the PSHA logic-tree, it should be recognized that the integration phase will involve building new models. There is an advantage in being aware of this from an early stage in the project. Specifically, the Project Plan should be developed to ensure that the evaluator experts have sufficient time to independently and objectively evaluate all data, methods, and models that currently exist within the larger technical community and then develop new models as necessary to capture the center, body, and range of technically defensible interpretations. The TI team members must fully assume the roles of technical integrators (and not proponents) in the development of new models. In a Level 3 project, in particular, members of the TI teams are likely to work in subgroups to develop various components of models. The team members must be given sufficient time to bring together these components into a coherent integrated distribution understood and endorsed by the entire team.

5.7   Demonstrating Capture of the Center, Body and Range

The core aim of the SSHAC process given in NUREG/CR-6372 is to represent the center, the body, and the range of the informed technical community (CBR of the ITC). As discussed in Section 3.1, for clarity, we have recast that SSHAC objective into two parts: (1) to consider the data, models, and methods of the larger technical community and (2) to represent the center, body, and range of technically defensible interpretations (the CBR of the TDI). The first objective is achieved through the evaluation part of the process and the second through the integration part. A question that can be asked is whether it can be demonstrated or proven that the process has effectively captured the CBR of the TDI, especially for a Level 3 or Level 4 projects where the investment of resources, time, and effort is considerable and the sponsors can justifiably expect some assurance that this objective has been met.

From the outset it is important to stress that there is no quantitative way to prove that the CBR of the TDI has been properly represented. One may hypothesize that some assurance could be provided that a stable estimate of the CBR of the TDI could be obtained by conducting two or more parallel projects with different teams of suitably selected experts. This could be done as a controlled experiment, in a similar fashion to the trial workshops that were conducted as part of the development of the original SSHAC guidelines (and documented in volume 2 of NUREG/CR-6372). The costs of conducting two real hazard assessments as Level 3 or 4 processes in parallel is likely to be prohibitively expensive. To cite Hanks et al. (2009): "*It is simply not possible to verify that the center, body, and range of the full ITC have been successfully captured without repeating the entire process of expert interaction with a different group of experts and perhaps different TI/TFIs as well. As a matter of practical reality, this has not occurred.*"

In the absence of any realistic means of proving so, confidence that the CBR of the TDI has been captured comes from the SSHAC assessment process itself and the confirmation from the PPRP that the project conformed to the requirements of the process. The essential steps outlined in Chapter 4 define the basic standard that is expected to lead to assurance that the goals of the SSHAC process have been met. Specific features of a Level 3 or Level 4 process that provide this assurance include the following:

- Careful selection of evaluator experts having the attributes described in Section 3.6.3.

- Appointment of a suitably qualified TI Lead or TFI, having the attributes and experience discussed in Sections 3.6.5 and 3.6.6.

- Identification of all applicable data and provision of an easily accessible and uniform database to all experts. It must be ensured that all modifications and additions are communicated to all the experts. The database development includes making full use of all local expertise by engaging those with first-hand knowledge of the site or region under study.

- Documented evaluation of the available data relative to their relevance, quality, and uncertainties for their use in the technical assessments for the study; Data Summary Sheets and Table Evaluation Tables (Section 4.10.2) or similar methods for documentation may be very useful in this respect.

- Identification of all potentially applicable models (and derivation of additional new models, if required, as discussed in Section 5.6 above) and the invitation of all relevant proponent experts to the appropriate workshop.

- Organization and occurrence of multiple expert interactions, encouraging challenge of the technical bases for assessments.

- Robust defense for the inclusion or exclusion of all models considered, and justification for the weights assigned to the selected models.

- Building of models and other expert assessments by the expert integrators through interactions among other experts on the panel in a Level 4 study and among the TI Team members in a Level 3 study.

- Testing of models, where appropriate (see Appendix Section B.4). This may not provide verification of models, but it can identify those models that are inconsistent with the available data.

- Sensitivity studies and feedback to evaluator experts.

- Integration of the models including developing all logic-tree branches and weights to represent the integrated distribution.

- Avoidance of artificial consensus through continuous technical challenge and defense, observed by the PPRP in the early stages of the project and partially conducted by the PPRP at Workshop 3. These communications also help avoid divergence that result from misunderstandings or exposure to different data sets or suites of candidates models.

- Peer review of process and of the justification for all technical assessments.

In calculating statistics of the hazard (e.g., the mean value), the weights on the logic-tree branches are treated as probabilities. This implies that the different models on the branches emerging from an individual node are assumed to be mutually exclusive and collectively exhaustive (Bommer and Scherbaum, 2008). In practice, meeting these so-called mutually exclusive and collectively exhaustive (MECE) criteria can be very difficult, particularly in terms of the models being mutually exclusive given that the datasets from which different ground-motion prediction equations are derived (e.g., will often overlap by including common records). The evaluators can, however, take this feature into account when assigning branch weights.

The imperative to capture the full range of the integrated distribution should not lead evaluator experts to include alternatives in their models only to convey the impression of broad capture of epistemic uncertainty. Only defensible models and parameter values should ever be included in the logic-tree. Assigning very small branch weights to models that the evaluator believes to be completely unsupported is not appropriate. On the other hand, there may be a genuine tendency for the distributions to be too narrow, partly because of the natural human tendency to underestimate epistemic uncertainty and the issues of anchoring (see Section 5.11). Another contributor to this tendency in some situations is that the evaluators may limit their assessments to published models, which by themselves may not span a sufficient range of outcomes to be considered as adequately capturing the full range of possibilities. An example of overcoming this difficulty is the 2008 USGS national seismic hazard maps for the western United States (Petersen et al., 2008). The study used three of the models (Boore and Atkinson, 2008; Campbell and Bozorgnia, 2008; Chiou and Youngs, 2008) produced by the NGA project (Power et al., 2008). These three predictive models were based on the same dataset, even though each model employed a different subset of the whole and give very similar median predictions for strike-slip earthquakes. For these reasons, they were considered to not to adequately capture the range of epistemic uncertainty associated with ground motions from future earthquakes in the western United States. Additional branches were added for lower and higher median motions in those magnitude-distance ranges where the equations are less well constrained by the data (Petersen et al., 2008).

5.8   Avoiding and Handling Excessively Large Logic-Trees

One consequence of earnestly seeking to capture the center, the body, and the range of the TDI is that the final logic-tree defining the input to the PSHA calculations can become very large in terms of the total number of branches and their combinations. The SSC and GMC sections of the logic-tree will each generally have several sequential sets of branches. In the SSC sub-project, these sets may include branches for source zones vs. smoothed seismicity (with subsequent branches for source geometry and smoothing distances, respectively), recurrence rates and maximum magnitudes, among many others. In the GMC subproject, the sets may include various models for the median motions, various options for adjusting these medians to the target V_s horizon (if needed), and alternative models for the aleatory variability (magnitude-dependent and constant, for example). If, as was the case for the PEGASOS project (Abrahamson et al., 2002), there is additionally a subproject for site response, another set of branches may follow for different interpretations of the geotechnical profile, different site-response analysis techniques, and different soil degradation models. If there are three or four branches for each of these models or parameters, there will be several tens of thousands of combinations for which hazard computations need to be performed.

In a Level 4 study, each member of each subproject expert panel develops an individual logic-tree and these are then all combined into a single integrated logic-tree; the total number of hazard calculations can become prohibitively large. In projects where this has happened, the hazard calculation team has sometimes simplified the logic-tree, using techniques that have become known as "pruning," "trimming," and "pinching" whereby branches are removed or merged to reduce the total number of hazard calculations. This simplifying of the logic-tree is performed taking care only to remove those elements that exert little influence on the final hazard results; however, in terms of process, it could be viewed as modifying the expert assessments and, thus, potentially undermining the evaluator experts' ownership of their models.

It is useful to clarify that with modern computing capacity, even the most complex logic-trees could be handled by obtaining access to sufficient computational power (although in the case of some commercial software, the license conditions can hamper attempts to benefit from parallel computing). The issue is really one of cost-benefit and whether such expense and inconvenience are warranted if many of the branches are increasing the number of calculations with negligible impact on the hazard estimates. At the same time, it should be recognized that regardless of the QA procedures adopted and followed, the chances of making errors in the calculations (and their going undetected) inevitably increases with the complexity of the hazard input model. The illusion of greater precision by virtue of very complex logic-trees may then actually result in loss of accuracy. In addition, hazard calculations that can be run in a shorter time facilitate greater exploration of sensitivity, which is a very valuable and important feature of any major PSHA: the work of the TI Team will always benefit from the evaluators and integrators being more 'hazard-informed'.

The need for calculational efficiency should not in any way limit the evaluator experts' honest assessment of the CBR of the TDI. At the same time, the experts should have adequate hazard feedback to understand which elements of their models are most important to hazard and should be the focus of their efforts. The purpose of hazard results presented at Workshops 1 and 3 is to provide feedback to the experts on the sensitivity of different elements of their logic-trees to the overall hazard results. This will enable the experts to make informed judgments about where to focus and prioritize their efforts but not to encourage the removal of branches. The experts are charged with demonstrating that consideration has been given to all legitimate models. Consequently, the documentation of their evaluation and integration process should include all the models considered, and those that are retained as branches in the logic-tree are assigned a finite weight. Those that are assessed to be unsupported by data or theory are assigned zero weight and not included in the logic-tree. If an expert has defensible reasons to exclude a particular model or parameter value, it is appropriate for the justification to be documented and the model or parameter value to not be included in the logic-tree.

One might ask why evaluator experts would concern themselves with the complexity of the logic-tree if their charge is to provide hazard input that captures the CBR of the TDI? The objective of effectively capturing the integrated distribution should remain paramount, and the evaluator expert should also be encouraged to avoid any *post facto* alterations to his or her logic-tree strictly for the sake of simplifying the hazard calculations. This could compromise the expert's ability to speak to that model and thus their sense of ownership of the model. At the same time, because the evaluator expert must assign weights to all branches, there is a clear and obvious benefit to an evaluator expert if the structure of logic-tree is kept simple, provided that important elements have not been neglected. The TI Lead in a Level 3 project and the TFI in a Level 4 project should seek a balance between these different considerations, without ever making the evaluator experts feel obliged to simplify the logic-tree if it does not concur with the expert's technical assessment.

Ultimately, the hazard calculation team may have to simplify the tree to be able to conduct the hazard calculations in a reasonable period of time and at reasonable cost. If that is the case, the project documentation should present both the experts' original models and the simplified models. The document must also show that the changes have not impacted the hazard results significantly. This may not be straightforward because of the correlations between different parts of the logic-tree. It is not sufficient to simply demonstrate that the influence of a particular branch on the SSC or GMC model is small, because the cumulative effects of any trimmed branches must be considered. The documentation should also clearly specify the output of the SSHAC process (the evaluator experts' full models). It must also be clear that the post-processing by the hazard calculation team is separate from and outside the SSHAC process.

An attractive alternative to simplifying logic-trees with large numbers of branches is to sample the branches in such a way that the hazard distribution is calculated to approximate (within an acceptable tolerance limit) results that would be obtained by performing the complete PSHA fully incorporating every single branch tip. Monte Carlo and other similar methods might be used for this purpose provided that it can be satisfactorily demonstrated that the mean hazard curve and the fractiles (typically the 5-, 15-, 50-, 85-, and 95-percentiles) are being adequately approximated. It would be beneficial in the future to develop guidelines and rules for sampling logic-trees in such a way that there is confidence that the mean hazard and the required hazard fractiles will be estimated to within acceptable limits of accuracy.

5.9   Interface Issues and their Resolution

A Level 3 or Level 4 PSHA will invariably be divided into subprojects primarily to reflect the fact that the SSC and GMC elements of the hazard input require different types of assessments and consequently involve different expertise. For example, the organization of a PSHA study as two separate subprojects for SSC and GMC was adopted for the Yucca Mountain PSHA (Stepp et al., 2001). In the PEGASOS project in Switzerland (Abrahamson et al., 2002), in which the hazard was calculated for four sites with different near-surface geology, there was an additional subproject for site response characterization. Because all the separate components of a PSHA study are ultimately combined into a single-logic-tree that defines all of the hazard calculations, it is essential that all the elements of the hazard model are compatible with one another and are defined using consistent parameter definitions.

If a hazard analysis is conducted using separate subprojects for SSC and GMC models, interface issues must be kept constantly in mind. If the projects run in parallel, then regular communication between the PTIs or PTFIs is to be encouraged; if in series (such as the case of the CEUS SSC and NGA-East projects), the PTI or PTFI from the first project can provide information on interface issues to the second project.

Figure 5-3 illustrates the main interface issues among the subprojects of a major PSHA study. The interface with the downstream engineering applications, discussed in Section 5.1, must also be considered. In Figure 5-3, the division between the ground motion and site response characterization subprojects is shown as a dashed line because these two may be merged into a single subproject, particularly when the hazard is to be calculated for a rock site. The right-hand side of the figure lists the key interface issues between each subproject, which are briefly discussed in the following paragraphs. Not illustrated in the figure are the internal consistency issues that can arise within the GMC or SSC characterization subprojects, such as the need to reconcile all the selected GMPEs to a common set of definitions of both the predicted and explanatory variables (e.g., Bommer et al., 2005).

The interface between the SSC and GMC sub-projects is related to parameter definitions and ranges. The recurrence rates for area and fault sources will be defined using a given magnitude scale, most commonly moment magnitude M_w. If any of the selected GMPEs use a different magnitude scale to quantify earthquake size, an adjustment—using a suitable empirical relationship—needs to be applied, and the standard deviation of that empirical relationship needs to be propagated into the standard deviation of the ground-motion prediction equation (Bommer et al., 2005).

If the SSC model specifies the style-of-faulting of the seismic sources (which it always will for faults and often will for areas), it is important that the GMPEs are able to account for the influence of this feature. Some ground-motion models, for example, only model the effect of some styles-of-faulting (e.g., only reverse or only reverse and strike-slip) and there are still some that do not consider style-of-faulting at all. For the latter case, if something is known about the composition of the dataset from which the model is derived, factors can be applied to model the effect of style-of-faulting (Bommer et al., 2003). Consideration also needs to be given to the fact that GMPEs use different ranges of rake angle to define style-of-faulting classes.

The SSC model will impose a maximum distance (R_{max}) of seismic sources from the site and the recurrence relationships will define the maximum magnitude (M_{max}) to be considered in the hazard integrations. Both of these limits are very likely to exceed the strict upper limits of applicability of many of the GMPEs as defined by the distributions of the datasets from which they are derived. Such extrapolations are almost inevitable in PSHA and the GMC evaluators need to specifically consider how well the models are likely to behave beyond their limits. The experts should also consider adjustments and additional logic-tree branches to reflect the greater epistemic uncertainty associated with such extrapolations. However, the degree of concern will clearly be determined by the extent of the extrapolation; in other words, by how far beyond their strict limits of applicability the GMPEs are to be applied. If major extrapolations are going to be required as a consequence of the SSC model, the GMC experts should be made aware of this at the earliest possible opportunity.

Subproject	Interface Issues
Seismic Source Characterization	Magnitude scale (M), M_{max}, Style-of-faulting classification, 3D rupture geometry, R_{max}
Ground Motion Characterization	V_{S30}, V_S kappa (κ), Single-station sigma (σ_{ss})
Site Response Characterization	
Hazard Calculations	Distance metrics (R), M_{min},

Figure 5-3. **Subproject Structure and Examples of Interface Issues Between the Various Sub-Projects.**

The above issues are related to the requirement that the GMC model must be consistent with the features of the SSC model. Interface issues also arise in the opposite direction if the GMPE requires parameters such as focal depth or the depth of embedment of fault ruptures. This will apply for models using hypocentral (R_{hyp}) or rupture (R_{rup}) distance metrics (Abrahamson and Shedlock, 1997) and models that include depth-to-top-of-rupture (Z_{TOR}), amongst others. Communication of the need for these parameters early in the project will ensure that the SSC model includes this information.

Within the GMC subproject, particularly if it is divided into separate projects for "rock" motions and site response, a key issue is clear definition of the geological horizon at which the rock/soil boundary occurs. The rock ground-motion must be defined consistently with the way the input to site response calculations will be performed. This boundary is typically defined using the average shear-wave velocity in the uppermost 30 meters (V_{s30}), but adjustments can also be made for the deeper shear-wave velocity (V_s) profile (e.g., Cotton et al., 2006). However, the application factors that only account for V_s differences to adjust the GMPE predictions to the reference site profile can lead to erroneous results if the damping of high-frequency motions by softer profiles is not also accounted for. This damping effect is often represented by the parameter kappa, κ (Anderson and Hough, 1984). Equivalent κ values for empirical GMPEs can be determined by inversion to allow this effect to be included in the adjustment as long as the κ value corresponding to the target horizon can also be estimated (Scherbaum et al., 2006).

Another interface issue between the rock motions and the site response is the approach to assigning aleatory variability. Ground-motion models are generally based on the ergodic assumption, whereby variation of motions over several locations is used as a surrogate for variations at a single site over the occurrence of several earthquakes (e.g., Anderson and Brune, 1999). This substitution of space for time can lead to inflated estimates of the ground-motion variability at a site as shown by analysis of multiple recordings from a single location (Atkinson, 2006). The difference between global and single-station sigma values reflects repeatable site effects that can be removed in a site-specific PSHA with good characterization of the geotechnical profile and rigorous site response calculation. If this component of variability is to be taken out of the hazard model, it becomes very important to identify where it will be subtracted (i.e., from the rock motions or from that site response calculations) so that it is taken out only once.

The final interface is that between the GMC subproject and the hazard calculations. Here there are two issues, the first being related to the distance metrics employed in each of the GMPEs and the geometry of the various earthquake scenarios used in the hazard calculations. For fault sources, it should be straightforward to calculate the distance using the definition for each scenario event. For area sources, the scenario events are typically treated as point sources, but this is not appropriate if the GMPEs use extended source metrics such as R_{rup} and R_{JB} (Abrahamson and Shedlock, 1997). This can be overcome by generating several virtual faults for each of the larger magnitude scenarios. These virtual faults can either have a random orientation or can be preferentially aligned if there is a basis for assigning more likely strike directions, although this obviously has an impact in terms of computational effort. The alternative is to apply empirical adjustments between different distance metrics (e.g., Scherbaum et al., 2004a), but this will lead to inflated assessments of modeling sigma values (Scherbaum et al., 2005).

The other interface issue between the GMC subproject and the hazard calculations is the selection of the minimum magnitude, M_{min}, to be considered in the hazard integrations. This value is chosen on the basis of excluding contributions from frequent small-magnitude earthquakes that can increase the hazard of PGA and high-frequency spectral accelerations but which produce ground motions having too little energy to be of engineering consequence (EPRI, 1989). Even if the non-damaging ground motions are excluded by using a CAV filter (EPRI, 2006) rather than excluding small-magnitude earthquakes, a value of M_{min} still needs to be specified. The GMC subproject needs to be aware of this because the value may be smaller than the strict lower limit of applicability of the GMPE, in which case the GMC evaluator experts need to consider the effects of the downward extrapolation of the predictive equations. Awareness of this issue may be heightened by the fact that some recent studies have shown that empirical GMPEs tend to overpredict ground-motion amplitudes at the lower magnitude limit of the dataset (Bommer et al., 2007; Atkinson and Morrison, 2009; Chiou et al., 2010).

The solution to avoiding interface problems is clear, early, and frequent communication between the subprojects. Clear written communication on these issues between the TI leads or TFIs is highly beneficial. The facilitation of such communication and the avoidance of problems at any of the interfaces discussed above is one of the key responsibilities of the PTI or PTFI (see

Section 5.2). Highlighting potential interface issues at an early stage is also a responsibility of the PPRP (see Section 3.6.8).

Organizing SSC and GMC workshops to be held together may also facilitate the early resolution of interface issues, particularly if these include joint sessions. If this is impractical, then the summaries from separate SSC and GMC workshops should be distributed to all project participants. In addition, the organization of one or more workshops specifically to discuss interface issues may be advisable.

5.10 Community-Based Versus Individual Sponsorship

If a regional model is unavailable, a single sponsor, most probably an energy utility or other site owner for a critical facility, will commission a site-specific PSHA. However, in any given country or region, seismic hazard assessments are likely to be required for several sites and also for many different types of facilities. In such situations, much benefit may be gained from a community-based assessment that implies sponsorship from a spectrum of end users rather than a single user group. The spectrum could include utilities, Federal agencies, and regulatory groups as well as private and public consortia.

A community of sponsors can pool resources to conduct simultaneous PSHA studies at several sites or else to develop regional models that can then be used in site-specific hazard assessments. SSHAC Level 3 and 4 processes can be equally applied to both site-specific and regional PSHA studies. Many elements of the input to site-specific PSHAs at different locations in a given region will be common. Examples are the historical and instrumental earthquake catalogue, the tectonic model for the region, and the ground-motion prediction equations.

Conducting SSHAC Level 3 or 4 studies to develop community-based SSC and GMC models for an entire region or country can offer many advantages. Duplication of work on the earthquake catalogue and the ground-motion models can be avoided. Better use can be made of the available resources, particularly the pool of expertise. If hazard assessments are required at several sites and the number of the available experts is limited, then this approach can also save considerable time if the alternative is several separate site-specific studies conducted in series. Because many elements of the input to each site-specific PSHA would be commonly defined, each study can be expected to produce more stable results because the bases for the hazard assessment are less likely to be challenged.

Another issue is the possibility of widely diverging hazard estimates at closely located sites (apart from differences arising from differences in surface geology) that would be reduced by the development of regional SSC and GMC models. One could argue that if independent Level 3 or 4 studies were conducted for closely spaced sites, each effectively capturing the CBR of the TDI, then significant differences should not arise. Some might go as far as to suggest that this would be a welcome mutual check, but such a position ignores all the significant economies (listed in the previous paragraph) lost in such an approach.

Such community-based approaches to develop SSC and GMC models for an entire region can be implemented by the end users pooling resources to conduct these projects that would produce commonly owned results. Individual site owners can then adopt the regional SSC and GMC models and apply them to site-specific assessments, refining them for local seismic sources and including site response modifications as required. Figure 5.4 schematically illustrates the relationship between regional studies and site-specific assessment. In such a scheme, if the regional studies are conducted as Level 3 or 4 processes, it may be acceptable to perform the local refinement at Level 2 (see Section 6.4).

Such a scheme is currently being implemented for nuclear sites in the central and eastern United States (CEUS) through two SSHAC Level 3 projects funded by several agencies. The CEUS SSC for Nuclear Facilities project and the NGA-East project (see discussion in Section 2.1) will respectively develop SSC and GMC models that will provide regional input to PSHA studies for nuclear facilities through the Central and Eastern United States. The advantage of multi-agency sponsorship is the mutual ownership of the models across a range of user groups. None of the groups within the community of sponsors is likely to question the efficacy of the hazard assessments because they will be co-owners. This should assist with expediting licensing and safety review cycles. Moreover, a degree of coherence and consistency among seismic hazard estimates for various types of critical facilities is desirable, especially in providing regulatory assurance.

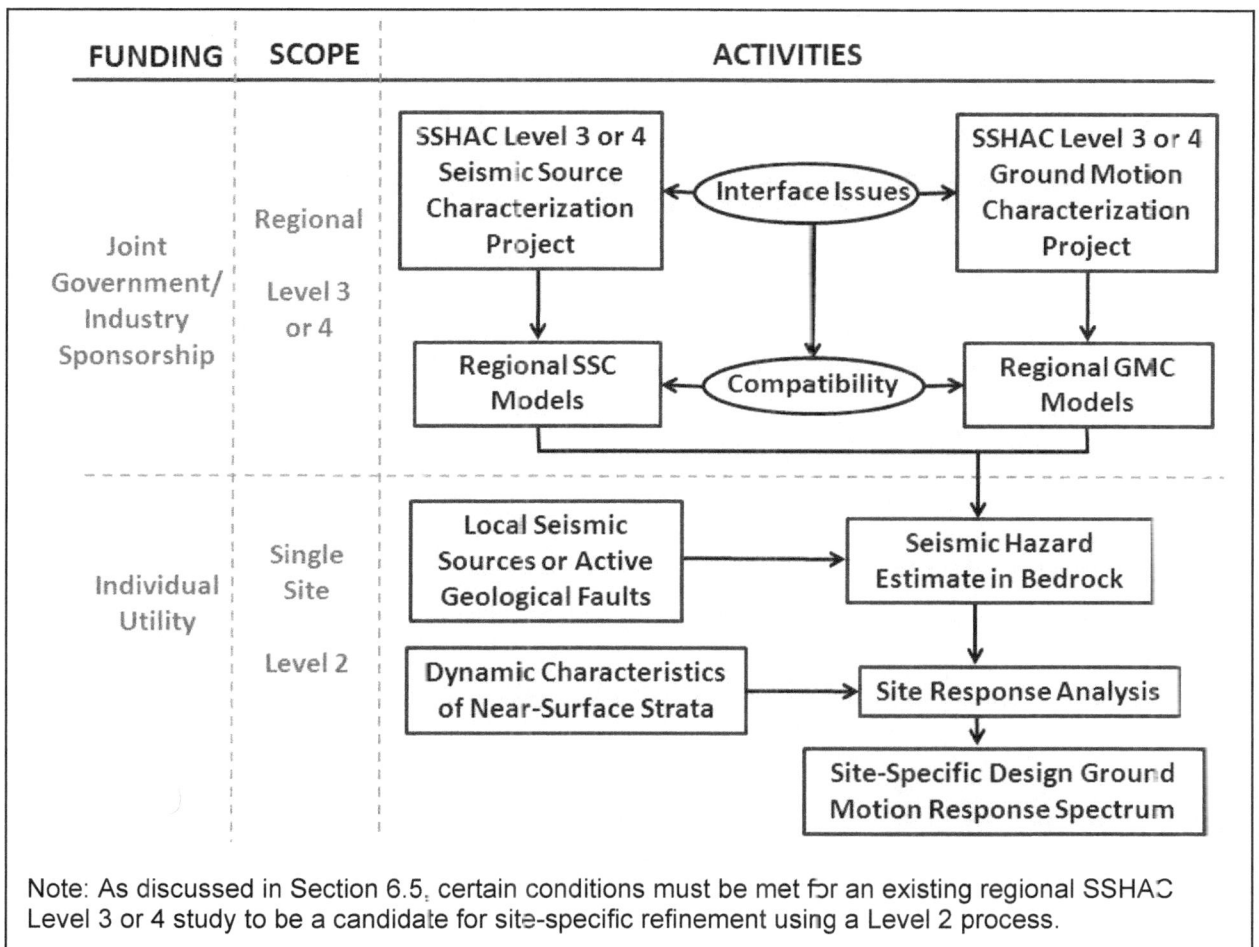

Note: As discussed in Section 6.5, certain conditions must be met for an existing regional SSHAC Level 3 or 4 study to be a candidate for site-specific refinement using a Level 2 process.

Figure 5-4. Example Organization of Level 3 or 4 Regional Studies Combined with Level 2 Site-Specific Refinements (Adapted from Bommer, 2010).

An alternative approach to the scheme illustrated in Figure 5-4 with single regional SSC and GMC projects is to conduct site-specific SSHAC Level 3 or 4 studies for several sites simultaneously as was done, for example, in the PEGASOS project. That study was carried out to perform comprehensive PSHA at four relatively closely spaced nuclear power plant sites in

Switzerland (Abrahamson et al., 2002). An advantage of the approach of producing regional community-based SSC and GMC models independently of site-specific PSHAs is that the regional study may be used by various end-users and at sites not identified at the time the study is conducted.

5.11 Cognitive Bias

Due to large uncertainties in our understanding and models of earthquake processes and the lack of empirical data to reduce this uncertainty, expert judgments are always required in seismic hazard analysis. In this field, advantage can be taken of a wealth of studies conducted regarding the use of experts to make estimates of quantities. There are several known problems that can plague expert assessments; most are not deliberate or intentional, but they must be countered. These problems are collectively referred to as cognitive bias.

Although there is a very extensive literature on the subject of cognitive bias, the treatment given here is necessarily brief and deliberately simplified. Consequently, it is unlikely to reflect the state-of-the-art in understanding of cognitive bias. However, for the purpose here, sophisticated philosophical discourses on the topic may be less useful than a succinct warning—repeated frequently during a PSHA project—about their existence and nature.

Examples of cognitive biases that are of clear relevance to conducting seismic hazard analyses include the following:

- Overconfidence: overestimating what is known (i.e., underestimating uncertainty).

- Anchoring: focusing on a specific number or model and not adjusting it sufficiently in light of new information.

- Availability: focusing on a specific, dramatic, or recent event; being inclined toward models that one is more familiar with or that one feels an affinity for because of knowing personally or by reputation the authors of a given model (or indeed, by being an author).

- Coherence/vividness: over-estimating the likelihood of an event because there is a 'good story.'

- Ignoring conditioning events: these are often unstated assumptions that influence the assessments that experts make.

It is important to note that (with the exception of motivational biases) the common cognitive biases, including those listed above, are inherent to all expert judgments and are not deliberate. This is simply the way that scientists and engineers commonly process information and offer technical judgments. Fortunately, studies have shown that the most effective way of countering cognitive biases is simply to make the experts aware that they exist and to encourage the experts to counter them. For example, the TI Lead or TFI can counter overconfidence by probing the limits of an expert's expression of uncertainty to ensure that the full range is being provided. Availability can be countered by asking the expert for the technical reasons that a particular model is preferred or that other models are considered less credible. The presence of ignored or unstated conditioning events can be brought to light by asking the expert for all assumptions that went into a particular expert assessment.

To capture the CBR of the TDI (Section 5.7), evaluator experts must be able to act as impartial and objective assessors of all available data, models, and methods (see Section 3.6.3). To achieve this, it is important to avoid cognitive bias in the assessments. Toward that end, the TI Lead/TFIs should discuss cognitive bias with the evaluator experts and make them aware that efforts will be devoted throughout the project to countering them. The PTI/PTFI or TFI/TI Leads should frequently remind project participants about the importance of avoiding cognitive bias (particularly in workshops and in working meetings where the experts are offering their judgments). The PPRP Chair should similarly remind the Peer Review Panel of the importance of being alert to cognitive bias. The expert interactions in Level 3 and 4 studies that specifically include technical challenge intended to reveal the genuine bases for experts' assessments is also a key component of countering cognitive bias.

Avoiding cognitive bias in the evaluations is a key responsibility of the TI Lead in Level 3 projects and of the TFI in Level 4 projects. Basic procedural tools, which are all closely related to one another, can be employed in Level 3 and 4 studies to militate against the influence of cognitive bias:

1.   Ensure that all data, models, and methods are made available to all the evaluators. This is the responsibility of the TI Lead/TFI, and the PPRP must ensure that it is achieved. Rather than selecting models that may be applicable, where availability and familiarity may introduce bias, all models should be compiled and only then should models be eliminated if judged to be unsuitable (e.g., Cotton et al., 2006; Bommer et al., 2010). In other words, all models must be considered applicable until a strong case can be made for their exclusion. This approach is fundamentally different from building up the logic-tree by including models considered to be applicable and acceptably reliable.

2.   When new models are developed specifically for the PSHA by members of the evaluation teams, other members of the team should be encouraged to subject these models to thorough review and technical challenge. Of equal importance, the model developers themselves must consciously relinquish the role of resource/proponent expert once the models are completed. Model developers must also be given sufficient time to be able to then objectively evaluate these models together with all others under consideration.

3.   Proponents of a wide range of relevant models should participate in workshops, including the authors of differing or controversial models. Part of the responsibility of the PPRP is to assist the TI Lead/TFI in identifying proponent experts (see Section 3.6.2) to present their models within a formal workshop.

4.   Expert evaluations must be subjected to technical challenge to ensure that there are transparent and defensible bases assessments. Technical bases should be provided for both for the weights assigned to the models included on the logic-tree branches and for the exclusion of other models. The TI Lead or TFI has responsibility for instigating and motivating such technical challenge within the workshops. The PPRP is responsible for ensuring that these challenges are made in the workshops and that the justifications for model weights and exclusions are adequately documented. As noted in Section 5.4, at Workshop #3 it is recommended that the PPRP participates directly in this technical challenge.

A particular type of bias that can arise is related to expectations about actual values whether these be maximum magnitudes, recurrence rates, ground-motion amplitudes, sigma values, or even the hazard itself. When an expert's assessment is influenced by a preconceived value, it is referred to as "anchoring" or the anchor-and-adjustment heuristic, which is illustrated when experts are provided with an initial estimate (anchor) that is then adjusted up and down (O'Hagan et al., 2006). The approach has been shown to often produce "biased judgments because people often make insufficient adjustment from the initial anchor" (O'Hagan et al., 2006). Anchoring usually occurs unintentionally, and it is important to be watchful for its appearance in a SSHAC evaluation process. The evaluator experts should not, consciously or unconsciously, be constrained by the assessments of other experts, the views of the TI Lead or TFI, or even their own previous assessments if new data or information has become available that should prompt them to update their judgments.

Another way that bias can be introduced through anchoring is if experts become reluctant to move significantly away from models and inputs developed early in the project for purposes of sensitivity analysis. For example, in a Level 3 or 4 study, preliminary SSC and GMC models are developed and hazard calculations conducted for purposes of sensitivity analyses and identifying the hazard-significant issues at Workshop #3 (Section 4.6.3). It is recommended that expert evaluators do not document in detail their preliminary models and their technical bases in the project report. The reason for this is if the expert has invested considerable effort and time in documenting early-stage assessments, they may become reluctant to update these assessments, even if the results or discussions with other expert evaluators prompt them to do so. This is one of the ways in which anchoring and availability biases can be related. Another example is when experts become anchored on a recent event, such as a published paper that has just become available in print. Other ways in which information can become an anchor through "availability" include being a current focus of attention in research or even in the news, being dramatic (unexpected but noticeable), being vivid (i.e., easily pictured), and being in some sense "official."

It is common for experts in PSHA to have worked on other projects in which the actual amplitude of calculated ground motions can have a profound effect on the manner in which the PSHA is used for design or safety purposes. For example, judgments are made about whether a site is a "high" or "low" hazard site based on comparisons of calculated ground motions at a given annual frequency of exceedance. Comparisons of this type are not useful during the development of a hazard model and can lead to potential bias in constructing the model such that it achieves a predetermined hazard result. To avoid early-stage hazard calculations becoming anchors to experts' assessments, it is suggested that any hazard results calculated early in the project for purposes of sensitivity analyses to define important issues (Workshop #1) be presented in normalized format (e.g., dividing all ground-motion values by the corresponding value for a well-defined base case and given annual exceedance frequency). Any format or presentation device is acceptable provided that it does not provide the evaluators with explicit associations of annual exceedance frequencies and ground-motion amplitudes. This is because this information could become an anchor and prevent sufficient updating of the experts' assessments when the data and analyses developed in the project would warrant significant changes to their earlier models. However, the normalization should not conceal the annual frequency of exceedance so that evaluators can see at what return period important changes occur.

In addition to preventing anchoring to early hazard results, the use of normalized hazard representations until the final assessments are made can offer another very important advantage because other factors may act as anchors on the hazard estimates. Significant examples of such anchors external to the PSHA project include the following:

- Existing seismic hazard estimates for the region where the site is located or for closely located sites.

- Previous seismic hazard studies for the same site; the issues associated with such comparisons are discussed in Section B.4 in Appendix B.

- If the sponsor has chosen a prequalified plant technology for which there is a specified ground-motion response spectrum (*e.g.*, Bommer et al., 2011), there may be a tendency to

110

influence the model such that the design response spectrum emerging from the PSHA does not exceed the certified design (or other reference) spectrum. In such cases, it becomes very important that the evaluator experts and hazard analysts are not aware of the numerical hazard results mid-project because inevitably the relationship between the spectral accelerations at certain annual exceedance frequencies and the ordinates of the design spectrum to which the plant technology has been pre-qualified could become a focus.

5.12 Quality Assurance

Embedded in the PSHA process described in NUREG/CR-6372, in these guidelines, and in ANSI/ANS-2.29-2008 (*American National Standard for Probabilistic Seismic Hazard Analysis*) is the concept of "participatory peer review," which is defined as both process and technical review of the PSHA starting at an early stage and continuing through the life of a project. This participatory peer review is a fundamental element in ensuring the quality of the resulting PSHA product. Both ANSI/ANS-2.29-2008 and ANSI/ANS-2.27-2008 (*American National Standard-Criteria for Investigation of Nuclear Facility Sites for Seismic Hazard Assessments*) were developed to be consistent with ANSI/ASME NQA-1-2008 *Quality Assurance Requirements for Nuclear Facility Applications.* Hence, following the guidance contained in these documents for a either a Level 3 or 4 assessment, NUREG/CR-6372, ANSI/ANS-2.29-2008 and ANSI/ANS-2.27-2008 will result in a study that satisfies the intent of national quality standards. For site-specific assessments that start with a Level 3 or 4 regional model and perform a Level 2 refinement for site-specific applications, compliance with specific requirements in ASME NQA-1-2008 Parts I and II will also be required.

Within the SSHAC hazard assessment framework, a traditional verification and validation (V&V) program is limited to specific numerical tools, such as the software used to perform the PSHA calculations. A quality or "cross-check" protocol may also be used to assure the accuracy of compiled tables, datasets, and other project products. However, it is not possible to apply a V&V program to the SSHAC process itself. Similarly, it does not make sense to impose a restriction on the use of data for cases where a formal quality assurance program for the collection of field data outside of the project cannot be verified (e.g., if a quality control program cannot be verified for a USGS or university dataset). The rejection of datasets in these cases could seriously diminish, instead of enhance, the process. This is because a key part of a SSHAC Level 3 or 4 process is the evaluation of data by the evaluator experts. Therefore, the evaluator experts are able to make an informed assessment of the quality of various datasets, whether or not those data were gathered within a formal quality program. This does not mean, however, that non-qualified data used in a SSHAC process can be considered qualified after their use in the process.

Within the SSHAC hazard assessment framework, the collection and evaluation of existing scientific information is performed with the aim of ascertaining the current state of knowledge regarding a specific issue. The majority of existing information that may be used in the conduct of a SSHAC Level 3 or 4 PSHA will have been published in some fashion previously. Moreover, the data, methods, and models considered and used will also undergo what effectively constitutes peer review by the TI team that is likely to be at least as rigorous as that conducted for journal publication. Thus, that information has been reviewed and "vetted" by the broad technical community. The systematic compilation of all pertinent information from the scientific literature (including specialized journals technical reports, conference proceedings) or other relevant sources of information (e.g., databases of scientific data, historical or archival documents) is a vital element in the conduct of a SSHAC PSHA study. In addition, in some cases, nontraditional types of data that may be beneficial to the project may be available. It is important that data not be dismissed without appropriate consideration, particularly in regions where data may be scarce.

Beyond the assurance of quality arising from that external scientific review process, a fundamental component of the SSHAC process is the evaluation of the data, models, and methods by the evaluator experts as a means of establishing the quality, relevance, technical basis, and uncertainties. Moreover, in the integration stage of the SSHAC assessment process, the TI team or evaluator experts build models and apply weights to elements of the model based on due consideration of the technical support for various models and methods proposed by the technical community. Therefore, it is the collective, informed judgment of the TI team (via the process of data evaluation and model integration) and the concurrence of the PPRP (via the participatory peer review process) as well as adherence to the national standards described above that ultimately leads to the assurance of quality in the process followed and in the products resulting from the SSHAC hazard assessment framework.

6. UPDATING: REPLACING AND REFINING PROBABILISTIC HAZARD ASSESSMENTS

The goal of a Senior Seismic Hazard Analysis Committee (SSHAC) process is to capture the center, body, and range of technically defensible interpretations at the particular point in time when the study is conducted. Therefore, a hazard study carried out using SSHAC guidelines is a "snapshot" in time. With the passage of time after a study, a number of changes can occur that might necessitate the consideration of updating or modifying an existing hazard study. These changes might be the gathering and analysis of new hazard-significant data, new models proposed by the larger technical community, or new methods for analyzing or interpreting data. In anticipation of these types of changes over time, it may be advisable to establish a fixed-term schedule for considering an update to a hazard study (e.g., every 10 years) or to put in place a process for evaluating the significance of new data, models, and methods as they become available. This chapter provides guidance on the issue of updating a hazard study conducted using either a SSHAC process or another process.

To provide a context for the subsequent discussion, the following definitions will be used relative to the issue of evaluating the applicability of an existing hazard study:

- Update or Updating. The process of first assessing whether or not an existing hazard study is acceptable for a current and specific use or requires replacement or refinement, followed by the completion of any necessary action.

- Replace. To completely set aside an existing hazard study and to develop a new study that will serve as the replacement to the previous study. It may be necessary to replace an existing hazard study if the previous study was not conducted in a proper SSHAC manner or if the data, models, and methods used are outdated and superseded such that the study cannot be modified to provide reliable results. Note that the mere passage of time is not necessarily a reason to replace an existing study.

- Refine. Starting with an existing regional hazard study conducted to acceptable SSHAC standards, the incorporation of site-specific information that may have a local influence in site-specific hazard. These refinements do not change the regional hazard model significantly unless new data, models, or methods locally would require such a change.

The guidance provided in this chapter is specific to nuclear facilities, which require the highest levels of regulatory assurance. The need to update a hazard study whose use might be for conventional buildings or other noncritical facilities could be quite different and is not addressed here.

It is also important to note that the recommendations made in this section, like those made elsewhere in the document, are intended to be recommendations based on experience gained from the implementation of past SSHAC projects. As stated in the original SSHAC document:

> "Note that our guidance is not intended to be "the only" or "the standard" methodology for PSHA to the exclusion of other approaches; there are other valid ways to perform a PSHA study. Likewise, our formulation should not be viewed as an attempt to "standardize" PSHA in the sense of freezing the science and technology that underlies a competent PSHA, thereby stifling innovation. Rather, our guidance is intended to represent SSHAC's opinion on the best current thinking on performing a valid PSHA." (NUREG/CR-6372, Budnitz et al., 1997, Executive Summary, reprinted as Appendix A)

Likewise, the information designated as recommendations made in this document are not requirements and are not intended to stifle innovation or discourage the use of other approaches. The recommendations are made in anticipation of the need for high levels of regulatory assurance required for nuclear facilities and the desire on the part of the user to employ methodologies that will have a high likelihood of demonstrating regulatory compliance. For example, recommendations are made for the SSHAC Level that would be used to replace or refine an existing hazard study. There is no prohibition against the sponsor of the study using a lower Level or non-SSHAC study with the added burden of having to convince the regulator that the data, models, and methods of the larger community have been considered and the center, body, and range (CBR) of the technically defensible interpretations (TDI) have been properly represented.

6.1 Use of the SSHAC Guidelines to Meet NRC Licensing Requirements

As discussed in Section 1.1, part of the motivation for this document is to provide additional guidance beyond the original SSHAC report for sponsors, practitioners, and reviewers of hazard studies conducted for regulatory and other purposes. By their nature, SSHAC Level 3 and 4 processes include a number of attributes that contribute to its success in dealing with contentious issues in a regulated environment. For example, processes include extensive participatory peer review throughout, and alternative and diverse technical viewpoints regarding controversial issues are highlighted and addressed through workshops and uncertainty treatment. As a result, higher-level SSHAC studies are inherently attractive for purposes of addressing licensing requirements and demonstrating regulatory compliance.

The principal geologic and seismic considerations for nuclear power plant site suitability are given in 10 CFR Part 100.23, *Geologic and Seismic Siting Criteria*. Reviews for Combined Operating License (COL) and Early Site Permit (ESP) applications have been conducted under 10 CFR Part 52, Subpart A and associated evaluation criteria from 10 CFR Part 100.23. Paragraph (d)(1), of 10 CFR 100.23 states, "*Determination of the Safe Shutdown Earthquake (SSE) Ground Motion requires that uncertainty inherent in estimates of the SSE be addressed through an appropriate analysis, such as a probabilistic seismic hazard analysis (PSHA)."* Regulatory guidance for the implementation of Part 100.23 and Appendix S to CFR Part 50 is provided in Regulatory Guide 1.208 (RG 1.208), "A Performance-Based Approach to Define the Site-Specific Earthquake Ground Motion."

Section B of RG 1.208 states that PSHA has been identified in 10 CFR 100.23 as a means to address the uncertainties in the determination of the SSE. Moreover, RG 1.208 recognizes that the nature of uncertainty and the appropriate approach to account for uncertainties depend on the tectonic setting of the site and on properly characterizing input parameters to the PSHA such as the seismic source characteristics, the recurrence of earthquakes within a seismic source, the maximum magnitude of earthquakes within a seismic source, and engineering estimation of earthquake ground motion through ground-motion prediction equations (i.e., attenuation relationships).

Probabilistic methodologies were developed specifically for nuclear power plant seismic hazard assessments in the central and eastern United States (CEUS). The experience gained by applying this methodology at nuclear facility sites, both reactor and nonreactor sites, throughout the United States served as the basis for development of the guidelines for conducting a PSHA captured in NUREG/CR-6372. RG 1.208 states: "*The guidelines detailed in NUREG/CR-6372 should be incorporated into the PSHA process to the extent possible* [emphasis added]. *These*

114

procedures provide a structured approach for decision making with respect to site-specific investigations. A PSHA provides a framework to address the uncertainties associated with the identification and characterization of seismic sources by incorporating multiple interpretations of seismological parameters." Hence, application of the SSHAC guidelines and procedures is fully consistent with U.S. Nuclear Regulatory Commission (NRC) regulations and, in fact, specified within RG 1.208 as an acceptable means to satisfy NRC regulatory requirements. That said, even the proper application of SSHAC approaches does not guarantee regulatory acceptance.

In the CEUS, the development of large regional seismic hazard studies have been conducted explicitly for nuclear power plants (i.e., the Electric Power Research Institute – Seismic Owners Group [EPRI-SOG] and Lawrence Livermore National Laboratory [LLNL] studies, see Section 2.1) and others have been conducted for purposes of the Building Code (i.e., United States Geological Survey [USGS] National Hazard maps). In the western United States, site-specific PSHAs have been conducted for the existing nuclear power plants. The seismic hazard component of Combined Operating License Applications that have been prepared in the past few years have been based on refinements to the existing EPRI-SOG study using a SSHAC Level 2 process. The recently completed SSHAC Level 3 CEUS SSC and ongoing Next Generation Attenuation Relationship for Eastern North America (NGA-East) studies (see project descriptions in Section 2.3) are intended to replace the EPRI-SOG and EPRI Ground Motion studies. It is expected that these studies will be recognized in updates to current regulatory guidance (i.e., RG 1.208) as acceptable for use in developing seismic design inputs for new nuclear power plants. A key reason for that acceptance is the fact that these studies are being conducted as SSHAC Level 3 studies with the direct participation of a number of stakeholders for nuclear facilities, including the NRC, nuclear utilities, and the U.S. Department of Energy (DOE). This is because such studies provide high levels of confidence that the larger technical community's views have been considered and that the CBR of the TDI have been properly represented.

6.2 The Need to Update Hazard Assessments

To every sponsor, analyst, and reviewer of hazard studies, the question arises as to how long such a study will be applicable and can be used and when a study has met the end of its useful life and needs to be replaced. The guidance discussed here will vary as a function of whether or not the previous study was conducted using a SSHAC process, the SSHAC Level of the previous study, and, of course, the presence and significance of new information that has become available since the previous study was conducted. It is important to remember that the overriding reason for considering the update of an existing hazard study is the same as the fundamental reason for employing the SSHAC process in the first place—the need to attain high levels of assurance that the larger technical community's data, models, and methods have been considered and the CBR of TDI have been represented. With the passage of time and, potentially, the development of new information, there will likely be a need to reassess whether that assurance continues to exist. Because both the sponsor and the regulator need confidence and assurance, the decision regarding whether or not to replace or refine a hazard study should be made in light of communication between the sponsor and regulator. As shown in the case of the Yucca Mountain Probabilistic Volcanic Hazard Analysis Update (PVHA-U), even if a licensee or applicant determines that an update (replacement or revision) is not needed, negotiations with the regulator may still result in one being carried out.

The decision regarding whether or not a hazard analysis requires some type of updating is an evaluation that requires careful consideration. Components of the updating evaluation require knowledge of the components of the PSHA (e.g., the SSC and GMC models and the data that drive the analyses), which in turn require knowledge of the hazard calculations and their

115

potential implications to their subsequent application. The evaluation process starts with the existing study and a detailed understanding of the data, models, and methods that were available at the time it was conducted. It will be against this backdrop that the new information will be compared and evaluated. A detailed inventory should be made of the regional and site-specific data, models, and methods that have become available since the existing study was conducted. Although new data may have become available, this does not automatically imply that the data will be important inputs to the hazard analysis. Typically, a very small subset of the total amount of earth science data developed in a region is applicable and significant with respect to hazard. For example, geologic and geophysical data are developed in abundance in some parts of the United States for purposes of oil and mineral exploration, but such data do not automatically have implications to the inputs to a PSHA.

The evaluation process uses this information to test whether it would lead to significant differences in the input models to the hazard analysis and to the hazard results. This is done by reviewing each individual component of the input models and making a judgment about whether and how the new information would lead to different inputs. The assessment includes evaluating the uncertainties and whether these would change with the new data.

For a reasoned decision to be made, criteria need to be established for what entails a "significant" change to the hazard analysis. For this assessment, it is recommended that two criteria be used: (1) an assessment should be made of whether or not the new information would lead to a change in the estimates of the CBR of the TDI in the major components of the model (e.g., SSC or GMC) and (2) an analysis should evaluate the magnitude of the change in the calculated hazard results and the significance to the subsequent use of the results. Either of these alone may indicate that an update is required. Clearly, changes in the calculated hazard results are the most diagnostic criteria that would inform a decision on whether or not to update a study. However, much of the credibility and confidence in a hazard study comes from the conclusion that it has appropriately captured the CBR of the TDI. Thus, even if a conclusion can be drawn that the calculated hazard results would not change significantly, large changes in the input models would need to be developed and documented in a SSHAC process to engender high levels of assurance. Simply put, the most important consideration in the decision process is confidence the model used in PSHA continues to have viable technical bases.

Assuming that new data, models, or methods lead to a change in the calculated hazard results, an assessment should be made of the significance of those changes to the intended use of the hazard results. Quantitative criteria for evaluating the significance of changes in calculated hazard results could be established mindful of their application[16]. For example, if hazard results will be used to establish design bases, some percentage difference in the design bases can be defined as the threshold between a significant and an insignificant change. Or if the hazard results will be used for a risk analysis, the threshold at which a change in hazard input will lead to a significant change in risk (e.g., defined as leading to noncompliance of a risk standard or as a percentage in risk) can be considered a measure of significance. The point of these types of assessments is to ensure that risk-informed significant changes in hazard are present and that these changes will motivate the need for an updating of a hazard study. Replacing an existing hazard study can be costly and time-consuming, hence the need to carefully evaluate the significance of new data, models, and methods. If a refinement is to be performed, it is not necessary for the people responsible for the original assessment to also carry out the refinement. However, the individuals performing the revision should satisfy the attributes described in Section 3.6.

---

[16] See Appendix G in the SSHAC report for an example of quantitative criteria for significance.

Closely related to the issue of hazard significance is the issue of the precision (or imprecision) of hazard estimates (e.g., McGuire, 2009). If a different group of equally qualified experts were given the same fundamental seismic data for a region (e.g., the same historical earthquake catalog, tectonic information, ground-motion data, site profile information, etc.), that group would derive a slightly different set of inputs and epistemic uncertainties. This would result in a slightly different estimate of mean hazard. Thus, any estimate of hazard has some associated imprecision regardless of how many experts are used in the assessment and how qualified they are. It is important to recognize this imprecision, attempt to quantify it, and evaluate the significance of possible future changes caused by new hypotheses or new data.

Two approaches reflect different ways of handling the issue of updating a hazard study: fixed-term updates and updates in response to new data, models, or methods. Each approach is discussed below.

6.2.1   Fixed-Term Updates

A reasonable way to address the anticipated changes that could motivate the need for an update to a hazard study is to schedule regular updates—or regular assessments of the need for an update—at some interval. Because the SSHAC guidelines have only been available relatively recently and nearly all of the SSHAC projects were conducted where no SSHAC study had been conducted previously, the average "shelf life" of a SSHAC hazard study cannot be estimated with high confidence. The only high-SSHAC Level study that has been updated with another such study is the SSHAC Level 4 PVHA conducted for Yucca Mountain (see Section 2.3). The PVHA was completely reassessed in light of new data, models, and methods about 12 years following the original study, resulting in very little change the assessed hazard results over that time period.

An example of fixed-term considerations of the need for updates is the process followed by the DOE. DOE Order 420.1B states that all natural phenomena hazards (NPH) assessments shall be reviewed every 10 years and evaluated for the need for an update:

> "3. REQUIREMENTS.
>> c. NPH Assessment.
>>> (4) An NPH assessment review must be conducted at least every 10 years and must include recommendations to DOE for updating the existing assessments based on significant changes found in methods or data. If no change is warranted from the earlier assessment, then this only needs to be documented."

An alternative to the fixed term for assessing the need for an update might be a fixed term for a required update regardless of the need. Perhaps with more SSHAC studies and experience, the requirement for an update on a fixed term would be feasible. But at the present time, there is not sufficient historical data to provide specific guidance or what an appropriate fixed term might be. For this reason, a fixed term approach is recommended for the evaluation of the need for an update. This alone can be a major undertaking for a regional study.

6.2.2   Updates in Response to New Data, Models, or Methods

An alternative to conducting assessments of the need for an update on a fixed-term schedule is to evaluate new information as it becomes available relative to its potential impact on the hazard. Clearly, this approach requires some level of constant or periodic vigilance with respect to new information being developed and the ability to assess the potential impact of that information on hazard. As an analog, nuclear power plant licensees are expected to evaluate the impact of new information or data on their license bases. This type of ongoing vigilance might be found in Federal agencies tasked with compiling and analyzing earth science information as part of their everyday operations. Importantly, the agency responsible also must have the knowledge and ability to evaluate the significance of new data, models, and methods to hazard. As discussed above in the introduction to Section 6.2, it also is important to consider the potential applications of the hazard study when making an evaluation of the significance of new data, models, and methods. Thus, the ongoing evaluations should anticipate the range of potential uses that might be made of the hazard results.

The advantage of the "continuous" updating approach is that it is more timely and responsive to changing data, models, and methods. As such, the approach allows a user to know at any time after a hazard analysis has been completed whether the study is still valid or needs updating. The approach is especially effective for hazard analyses within a rapidly changing environment. For example, methods for conducting probabilistic tsunami hazard analysis have evolved rapidly in the past 10 years as have the approaches to assessing PVHAs. As the methods evolve— usually incorporating increasing physical realism—the potential for significant changes in the existing hazard studies increases. As another example, increasing sophistication in the use of paleoseismic data in the CEUS has led to improved spatial and temporal models for earthquake occurrence that can motivate the need to evaluate the significance of potential changes to existing seismic hazard models. A continuous approach to evaluating the significance of these changes would provide valuable information for those who plan to use the existing hazard studies.

Key disadvantages of the "continuous" updating approach is that it would require a commitment on the part of a particular agency or funding partners to provide the necessary sustained resources (funding and availability of key personnel) for this approach to be successful. There may also be erroneous perceptions that continual (or periodic reviews with high frequency) reviews could undermine regulatory stability because new findings could lead to significant changes in the calculated hazard results with each review. These concerns should be allayed over time because it is expected that most new information will be shown to not lead to a significant change in hazard.

6.3   Replacement of Previous Hazard Assessments

As a context to the discussions in the remaining sections of this chapter, Table 6-1 presents the recommendations regarding the need to update an existing study and its SSHAC Level as a function of the nature of the existing study (i.e., regional or site-specific), the viability of the existing study, the nature of the needed study, and the SSHAC Level. These conditions and recommendations are discussed further below with particular focus on the cases requiring replacement or refinement.

Table 6-1. Recommendations Regarding Updating Hazard Assessments for Nuclear Facilities

Existing Study	Condition of Existing Study	Hazard Assessment Needed	Recommendation	SSHAC Level for New Study
No study, or previous studies conducted at lower SSHAC Levels (2 or 1), or non-SSHAC studies	Not adequate for nuclear/critical facilities	Regional and/or site-specific	Conduct new study	3 or 4
Regional or site-specific	Not viable**	Regional and/or site-specific	Replace existing study	3 or 4
Regional or site-specific	Viable	Site-specific	Refine regional study locally consistent with RG 1.208 and ANSI/ANS-2.27 / 2.29 2008	2, 3, or 4
Site-specific (one or more sites), no regional	Viable	Regional	Use site-specific studies to assist development of regional models	3 or 4
Site-specific (one or more sites), no regional	Not Viable	Regional	Conduct new study	3 or 4

** "Viable" is defined as: (1) based on a consideration of data, models, and methods in the larger technical community, and (2) representative of the center, body, and range of technically defensible interpretations.

Assuming that an evaluation process is carried out (as described above in Section 6.2) to determine the viability of existing hazard assessments, this section describes the situation in which a decision is made to replace the previous hazard assessments. First, a distinction is made between a regional and a site-specific hazard assessment. A regional hazard assessment is designed to provide hazard results over a region (e.g., the CEUS), and the results are commonly "mapped" to show the spatial variation. The level of detail in the inputs to the analysis is consistent with the needs of a regional study. A site-specific hazard analysis provides hazard results at a point or very small local area and is typically used for purposes of design or risk evaluation at that site. The spatial extent of the inputs is limited to those that would affect the site hazard, but the level of detail in the inputs is typically greater than the regional study. This is because local characteristics, such as the exact location of a seismic source boundary, can affect the site-specific hazard results but would likely not affect the regional hazard results significantly.

The fundamental criterion for deciding whether an existing hazard study needs to be replaced is whether it is still technically viable. "Viable" is defined to mean (1) the study properly and completely considers the data, models, and methods of the larger technical community and (2) it is representative of the CBR of the TDI. This is, of course, the current technical community and interpretations. If no previous hazard study exists or previous studies were not conducted using a SSHAC Level 3 or 4 approach, then there is a clear need to conduct a new hazard analysis. Assuming that a previous regional hazard study was conducted, an evaluation should be made as described in Section 6.2 to determine the need for updating. If very significant differences can be identified in the inputs to the hazard analysis as well as in the calculated hazard results from those that are perceived to currently exist, then the previous study will need to be replaced.

Clearly, the exploratory studies carried out to evaluate whether or not a hazard study needs to be replaced must be done in an expedited manner and are not the same as carrying out a complete study. Therefore, the assessment of whether the existing study is viable (as well as the potential changes in the calculated hazard results) must be based on limited evaluations and expert judgment.

Assuming that a site-specific study exists and that a new site-specific hazard estimate is needed, then the evaluation process would be the same as described for replacing a regional study with a new regional study. The new data, models, and methods that have become available since the previous study was conducted should be identified and their implications assessed relative to the previous study.

Assuming that a regional hazard study exists without a site-specific study and that a site-specific hazard estimate is needed, then an evaluation should be made of the viability of the regional study. The advantage of conducting a site-specific study at a location with an existing regional study is that the scope can be reduced and limited to local refinements as needed. If the existing regional study is not found to be viable, then the scope of the new site-specific study will need to be bolstered to include both the regional and local components.

Assuming that one or more site-specific studies exist without a regional study and that a regional hazard assessment is needed, the site-specific studies can be evaluated for their viability. The assessments made for the site-specific studies can provide "local control" for the regional study and potentially help limit the scope of the new regional study that will need to be carried out.

The Standard ANSI/ANS-2.29-2008 Probabilistic Seismic Hazard Analysis is cited by Department of Energy guidance as an acceptable approach to planning and conducting a

PSHA. With regard to the issue of updating an existing hazard analysis, the Standard specifies a series of high level requirements for assessing whether or not an existing hazard study is adequate for continued use without modification. Table 3 in Section 4.1 of ANSI/ANS 2.29 contains the High Level Requirements, which are outlined in Table 6.1 along with the questions that can be asked to determine whether or not the high level requirements have been met by the existing hazard study. It should be noted when referencing ANSI/ANS 2.29, commercial nuclear power reactors are considered to be SDC 5.

As seen by the various requirements given in the Standard, the continued use of an existing hazard study is only merited if the study includes a consideration of the current data, models, and methods of the technical community, and it properly accounts for aleatory and epistemic uncertainties. These criteria are essentially the same as those identified above in the definition of "viability."

6.4 SSHAC Level for Refinement of Site-Specific Assessments

Given that a viable regional hazard assessment exists, a new site-specific assessment can be made by refining the regional model. Such refinements can be made using SSHAC Level 2, 3, or 4 processes and should include a consideration of the databases that will be developed as part of licensing activities for a nuclear facility. For example, RG 1.208 (p.4) requires the development of an up-to-date site-specific earth science database that can be used for a PSHA and, in turn, the development of the design ground motions. As discussed in the RG, the studies conducted to support the database are designed to provide increasing levels of specificity moving from the site region to the site location. ANSI/ANS-2.27-2008, "Criteria for Investigations of Nuclear Facility Sites for Seismic Hazard Assessments," also presents criteria for site-specific investigations for purposes of ground motion and fault rupture hazard analysis.

The site-specific earth science database can provide the technical basis for refinements made to the regional hazard model. For example, the existing regional hazard model may place the site within a particular seismic source having a set of seismic source characteristics (e.g., m_{max} distribution, spatially varying recurrence rates) surrounded potentially by other sources each having their own characteristics. The site-specific earth science database should then be used to refine the model. For example, the existence and position of seismic source boundaries can be evaluated in light of more specific tectonic data, an updated earthquake catalog can be used to revise recurrence rates and their spatial variation, local paleoseismic datasets may be included to characterize local seismic sources, and local tectonic features can be evaluated for their seismogenic potential, as needed.

In the refinement process, the existing regional model should be studied carefully to gain a thorough understanding of the manner in which it considers the larger technical community and captures the CBR of the TDI on a regional basis. Then, in the refinement process, care should be taken to maintain the fundamental elements of the regional model unless local data dictate that changes are needed and appropriate. Typically, site-specific refinements to a regional model involve additional complexity and specificity where it previously did not exist. For example, adding a local fault or local tectonic feature (with associated probability of activity) is a typical refinement. But, if the data gathered as part of the site-specific studies so indicate, the refinements may also involve local modifications to the regional model, such as moving source boundaries or reassessing the magnitude and location of earthquakes in the site region—and the possible consequences to source characteristics such as maximum magnitude estimates. All changes of this type should be justified by additional local studies and development of the site-specific earth science database. All changes and their technical bases should be thoroughly documented.

6.5 Maintenance and Evaluation of Databases Between Hazard Assessments

Earth science data that are potentially important to hazard assessments are developed on a continuous basis regardless of whether or not a hazard assessment is conducted. For example, regional and global seismicity data are constantly being gathered as are geodetic data in many areas. Geologic and geophysical studies are commissioned for a variety of applications, and ground-motion data are being gathered worldwide that have potential significance to a hazard assessment. In the past, the compilation of significant data for a hazard analysis was conducted primarily during the course of power plant licensing, major studies like the EPRI-SOG and LLNL studies, or on a site-specific basis for major facilities such as Yucca Mountain. Unfortunately, the hiatus in such licensing activities during the period 1985 to 2005 meant very little was done to update and compile seismic-hazard related datasets. The USGS National Seismic Hazard Mapping Program (NSHMP) has a process for reviewing and compiling relevant data for updates to the maps about every 5 years. However, the focus of the USGS NSHMP is on a range of annual exceedance frequencies quite different from those of interest for nuclear facilities. Hence, the data that are compiled and the hazard assessments may differ from that appropriate for use at nuclear facilities.

The recommencement of licensing activities in the early 2000s led to local updates of the available data for use in revising the existing regional hazard models developed over 20 years prior. The ongoing CEUS SSC project and NGA-East projects are the first community-based regional higher-level SSHAC studies, and the systematic compilation and evaluation of available data, models, and methods are a key part of their activities as SSHAC Level 3 projects. In both studies, as well, the development of a uniform and accessible database is a key deliverable. Once completed, the existence of those databases could potentially provide unique opportunities to consider more systematic approaches to maintaining and evaluating databases in the future. That is, rather than rely on licensing activities to provide the mechanism for keeping the database current, an effort could be initiated to provide a continuous updating of the database. Consistent with the database having particular applicability to hazard analyses, the data updating effort should include an evaluation of the data as they become available for their quality and applicability to hazard analyses. The CEUS SSC and NGA-East projects are jointly sponsored by both the public and private sectors based on the premise that a community-based effort would have broad-based support and credibility. It is suggested that a similar community-based initiative—sponsored by all of the groups that would benefit—be considered for a future database compilation and evaluation effort. It is recognized this would require a stable repository of information over a long period of time.

7. SUMMARY

In light of the discussions given in this document, this section summarizes the key conclusions and recommendations. These are provided as a means of calling attention to important issues but are not intended to downplay any of the other conclusions and recommendations made throughout the document. The reader is encouraged to refer to the text to provide the appropriate context for the points made in this chapter.

- Considering the data, models, and methods from the larger technical community (evaluation) and capturing the center, body, and range of technically defensible interpretations (integration) remain the core objectives of the Senior Seismic Hazard Analysis Committee (SSHAC) process. Although these objectives are applicable to all SSHAC Levels, the higher SSHAC Levels provide higher confidence that the objectives have been reached. Chapter 4 provides the minimum essential steps that must be included in any SSHAC Level 3 or 4 study.

- The SSHAC process has now been implemented in a number of SSHAC Level 3 and 4 studies (given in Chapter 2), thus providing a basis for recommendations in terms of detailed implementation guidance. The lessons learned from past projects serve to strengthen the recommendations while acknowledging that continued refinements to SSHAC processes will continue to be made in the future. Unlike the original SSHAC document that focused on Level 4 studies, this report provides detailed implementation guidance for Level 3 studies as well.

- While not providing a prescriptive formula, this document does include a description of the criteria and selection process that is recommended for arriving at the appropriate SSHAC Level for a given project.

- Whereas there has been a perception that the most significant increase in rigor, cost, and duration occurs in moving from a Level 3 to a Level 4 study, the major jump is actually between Level 2 and Level 3. From the regulatory perspective of the NRC, there is no essential difference between Level 3 and Level 4 studies, and throughout these guidelines they are considered as parallel and equally valid options.

- Defining roles and associated responsibilities is important. Roles discussed in Chapter 3 include the Participatory Peer Review Panel (PPRP), the Project Technical Integrator (PTI), Technical Integrator (TI) Leads, Technical Facilitator Integrator (TFI), sponsors, etc. To ensure that the roles are respected throughout the project, it is important to set expectations at the outset in the statements of work for each participant.

- The development of a comprehensive Project Plan, including all activities and their schedule, is an essential step in a SSHAC project. The Project Plan sets expectations for all project participants as well as those who are observing or reviewing the study. The Project Plan also should show how the project will be consistent with the essential process steps given in Chapter 4 of this document.

- The 11 essential steps required to claim that a Level 3 or 4 study has been carried out are defined in Chapter 4 and include a minimum of 3 formal workshops focused on specific areas, conducted according to clear rules, and observed throughout by the PPRP.

- Whereas the SSHAC guidelines assigned the PPRP the role of observers at these workshops, these implementation guidelines recommend that at Workshop #3 the PPRP be allowed to engage and directly question the experts responsible for the evaluation and integration. This may prevent significant technical issues arising during the documentation phase.

- Experience during SSHAC Level 3 and 4 projects shows that the role of Project TFI (PTFI) or Project TI (PTI) can be very beneficial. The PTFI/PTI is responsible for the technical aspects of the project and is responsible for resolving interface issues between the components of the project. The PTFI/PTI is also the point of contact between those responsible for the technical aspects and the Project Manager, who is responsible for maintaining the scope, schedule, and budget.

- Complete documentation of data, models, and methods is key to a successful project. The goal is to provide the reader with a clear understanding of the technical bases for all assessments including the associated uncertainties. Data tables to record the data evaluation process are an example of a documentation tool that has been shown to be successful.

- Because of a paucity of empirical data, expert judgment continues to be important in the model-building process. The experience gained in recognizing and countering known cognitive biases can be brought to bear in the model development and expert evaluation process. A specific recommendation to assist in avoiding the perils of anchoring is that until the project enters the model-building integration phase, the project participants should only be shown interim hazard results in a normalized format so that attention is not focused on the implied design ground-motion levels.

- The Hazard Input Document (HID) has been shown to be an effective tool to capture the essence of the technical assessments for use in subsequent hazard calculations. As such, the HID is also a tool for the TI team or evaluator experts to verify the accuracy of the assessments that are being provided to the hazard analysts.

- Community-based hazard assessments—those that shared sponsorship across the public and private sectors—have distinct advantages, and we now have project experience showing this to be the case. The experience from ongoing projects shows that the needs of a diverse group of sponsors can be met with component costs to each agency that are less than they would be in a series of separate studies.

- Testing of PSHA should be conducted with caution, and conclusive validation is not possible given the rare nature of the important hazards. It is possible, however, to partially test components of the models, and different tests are applicable for different levels of probability.

- Based on consideration of the balance between costs and the need for regulatory assurance, new hazard studies conducted for purposes of the licensing of new nuclear facilities should be conducted to SSHAC Level 3 or 4. New site-specific studies may be site-specific conducted as a Level 2 refinement in cases where a SSHAC-based regional study exists. Chapter 6 provides guidance on the need for updating and the manner in which existing studies should be updated. Recommendations are made in light of the viability of existing studies as well as the needs of the planned study.

- Despite the focus on seismic hazards in this document, the methodology can be applied to other natural hazards as well. These hazards include tsunami, fault displacement, volcanism, flooding, and liquefaction. Appendix B of this document provides an overview of probabilistic hazard analysis for earthquake ground shaking and brief explanations of how the same framework can be adapted for other geological hazards.

- The guidance provided is not specific to the United States but can be applied globally; indeed, these guidelines reflect lessons learned from implementation of the SSHAC process in Canada, Switzerland, and South Africa as well as the United States.

- The SSHAC implementation recommendations made in this document are intended to provide a stable framework for future studies to ensure that the process is consistent and transparent. By specifying the essential *process* aspects of the study, it is assumed that future projects can focus on the technical assessments in the hazard analysis. It is also recognized that future studies will likely result in enhancements and refinements to the basic elements presented here. This is expected and encouraged.

8. REFERENCES

Abrahamson, N.A. (2000). State of the practice of seismic hazard assessment. *Proceedings of GeoEng2000*, Melbourne, Australia, vol. 1, 659-685.

Abrahamson, N., G. Atkinson, D. Boore, Y. Bozorgnia, K. Campbell, B. Chiou, I.M. Idriss, W. Silva and R. Youngs (2008). Comparisons of the NGA ground-motion relations. *Earthquake Spectra* **24**(1), 45-66.

Abrahamson, N.A., P. Birkhauser, M. Koller, D. Mayer-Rosa, P. Smit, C. Sprecher, S. Tinic and R. Graf (2002). PEGASOS – a comprehensive probabilistic seismic hazard assessment for nuclear power plants in Switzerland. *12th European Conference on Earthquake Engineering*, London, Paper No. 633.

Abrahamson, N.A. and J.J. Bommer (2005). Probability and uncertainty in seismic hazard analysis. *Earthquake Spectra* **21**(2), 603-607.

Abrahamson, N.A. and K.M. Shedlock (1997). Overview. *Seismological Research Letters* **68**(1), 9-23.

Akkar, S. and J.J. Bommer (2006). Influence of long-period filter cut-off on elastic spectral displacements. *Earthquake Engineering & Structural Dynamics* **35**(9), 1145-1165.

Akkar, S., Ö. Kale, E. Yenier and J.J. Bommer (2011). The high-frequency limit of usable response spectral ordinates from filtered analogue and digital strong-motion accelerograms. *Earthquake Engineering & Structural Dynamics, 40(12), 1387-1401.*

Al-Atik, L., N.A. Abrahamson, J.J. Bommer, F. Scherbaum, F. Cotton and N. Kuehn (2010). The variability of ground-motion prediction models and its components. *Seismological Research Letters* **81**(5), 783-793.

Albarello, D. and V. D'Amico (2005). Validation of intensity attenuation relationship. *Bulletin of the Seismological Society of America* **95**(2), 719-724.

Aldama-Bustos, G., J.J. Bommer, C.H. Fenton and P.J. Stafford (2009). Probabilistic seismic hazard analysis for rock sites in the cities of Abu Dhabi, Dubai and Ra's Al Khaymah, United Arab Emirates. *Georisk* **3**(1), 1-29.

Allen, T.I. and D.J. Wald (2009). Evaluation of ground-motion modeling techniques for use in Global ShakeMap: A critique of instrumental ground-motion prediction equations, peak ground motion to macroseismic intensity conversions, and macroseismic intensity predictions in different tectonic settings. *USGS Open File Report 2009-1047*, US Geological Survey, Reston, Virginia.

Anderson, J.G. (2010). Engineering seismology: directions in probabilistic seismic hazard analysis. *Proceedings of 7th International Conference on Urban Earthquake Engineering (7CUEE) and 5th International Conference on Earthquake Engineering (5ICEE)*, March 3-5, Tokyo Institute of Technology, Japan.

Anderson, J.G. and J.N. Brune (1999). Probabilistic seismic hazard assessment without the ergodic assumption. *Seismological Research Letters* **70**(1), 19-28.

Anderson, J.G. and S.E. Hough (1984). A model for the shape of the Fourier amplitude spectrum of acceleration at high frequencies. *Bulletin of the Seismological Society of America* **74**(5), 1969-1993.

Andrews, D.J., T.C Hanks and J.W. Whitney (2007). Physical limits on ground motions at Yucca Mountain. *Bulletin of the Seismological Society of America* **97**(6), 1771-1732.

ANSI/ANS-2.26-2004, Categorization of Nuclear Facility Structures, Systems, and Components for Seismic Design, American Nuclear Society and American National Standards Institute National Standard.

ANSI/ANS-2.27-2008, Criteria for Investigations of Nuclear Facility Sites for Seismic Hazard Assessments, American Nuclear Society and American National Standards Institute National Standard.

ANSI/ANS-2.29-2008. Probabilistic Seismic Hazard Analysis. American Nuclear Society and American National Standards Institute National Standard.

ANSI/ASME NQA-1-2008-1, Quality Assurance Requirements for Nuclear Facility Applications. American National Standards Institute and American Society of Mechanical Engineers, National Standard.

Arias, A. (1970). A measure of earthquake intensity. *In*: Seismic Design for Nuclear Power Plants, R. Hansen *ed.*, MIT Press, Cambridge, Massachusetts, 438-483.

ASCE (2004). Seismic design criteria for structures, systems, and components in nuclear facilities and commentary. *ASCE Standard 43-05*, American Society of Civil Engineers.

Atkinson, G.M. (2006). Single-station sigma. *Bulletin of the Seismological Society of America* **96**(2), 446-455.

Atkinson, G.M. (2008). Ground-motion prediction equations for Eastern North America from a referenced empirical approach: Implications for epistemic uncertainty. *Bulletin of the Seismological Society of America* **98**(3), 1304-1318.

Atkinson, G.M. and D.M. Boore (2006). Earthquake ground-motion prediction equations for Eastern North America. *Bulletin of the Seismological Society of America* **96**(6), 2181-2205.

Atkinson, G.M. and S.I. Kaka (2007). Relationship between felt intensity and instrumental ground motions. *Bulletin of the Seismological Society of America* **97**(2), 497-510.

Atkinson, G.M. and M. Morrison (2009). Observations on regional variability in ground-motion amplitude for small-to-moderate magnitude earthquakes in North America. *Bulletin of the Seismological Society of America* **99** (4), 2393-2409.

Baker, J.W. and C.A. Cornell (2006). Spectral shape, epsilon and record selection. *Earthquake Engineering and Structural Dynamics* **35**, 1077-1095.

Bakun, W.H. and C.M. Wentworth (1997). Estimating earthquake location and magnitude from seismic intensity data. *Bulletin of the Seismological Society of America* **87**(6), 1502-1521.

Bazzurro, P. and C.A. Cornell (1999). Disaggregation of seismic hazard. *Bulletin of the Seismological Society of America* **89**(2), 501-220.

Beauval, C., P.-Y. Bard, S. Hainzl and P. Guéguen (2008). Can strong-motion observations be used to constrain probabilistic seismic-hazard estimates? *Bulletin of the Seismological Society of America* **98**(2), 509-520.

Bedford, T. and R. Cooke (2001). *Probabilistic risk analysis: foundations and methods.* Cambridge University Press, Cambridge, UK.

Bernreuter, D.L. and C. Minichino (1982), Seismic hazard analysis overview and executive summary, US Nuclear Regulatory Commission NUREG/CR-1582-Vol.1; UCRL-53030.

Bernreuter, D.L., J.B. Savy, R.W. Mensing, J.C. Chen, and B C. Davis (1999). "Seismic Hazard Characterization of 69 Nuclear Plant Sites East of the Rocky Mountains," NUREG/CR-5250, Volumes 1–8, U.S. Nuclear Regulatory Commission, Washington, DC.

Beyer, K. and J.J. Bommer (2006). Relationships between median values and aleatory variabilities for different definitions of the horizontal component of motion. *Bulletin of the Seismological Society of America* **94**(4A), 1512-1522. *Erratum* 2007, **97**(5), 1769.

Bird, J.F. and J.J. Bommer (2004). Earthquake losses due to ground failure. *Engineering Geology* **75**(2), 147-179.

Blaser, L., F. Krüger, M. Ohrnberger and F. Scherbaum (2010). Scaling relations of earthquake source parameter estimates with special focus on subduction environment. *Bulletin of the Seismological Society of America* **100**(6), 2914-2926.

Bommer, J.J. (2010). Seismic hazard assessment for nuclear power plant sites in the UK: challenges and possibilities. *Nuclear Future* **6**(3), 164-170.

Bommer, J.J. and N.A. Abrahamson (2006). Why do modern probabilistic seismic hazard analyses lead to increased hazard estimates? *Bulletin of Seismological Society of America* **96**(6), 1967-1977.

Bommer, J.J., N.A. Abrahamson, F.O. Strasser, A. Pecker, P-Y. Bard, H. Bungum. F. Cotton, D. Faeh, F. Sabetta, F. Scherbaum and J. Studer (2004). The challenge of defining the upper limits on earthquake ground motions. *Seismological Research Letters* **75**(1), 82-95.

Bommer, J.J. and D.M. Boore (2004). Engineering Seismology. *In*: Encyclopedia of Geology, Academic Press, vol. 1, pp.499-514.

Bommer, J.J., J. Douglas and F.O. Strasser (2003). Style-of-faulting in ground motion prediction equations. *Bulletin of Earthquake Engineering* **1**(2), 171-203.

Bommer, J.J., J. Douglas, F. Scherbaum, F. Cotton, H. Bungum and D. Fäh (2010). On the selection of ground-motion prediction equations for seismic hazard analysis. *Seismological Research Letters,* **81**(5), 794-801.

Bommer, J.J., J. Hancock and J.E. Alarcón (2006). Correlations between duration and number of cycles of earthquake ground motion. *Soil Dynamics and Earthquake Engineering,* **26**(1), 1-13.

Bommer, J.J. and A. Martinez-Pereira (1999). The effective duration of earthquake strong motion. *Journal of Earthquake Engineering* **3**, 2, 127-172.

Bommer, J.J., M. Papaspiliou & W. Price (2011). Earthquake response spectra for seismic design of nuclear power plants in the UK. Nuclear Engineering & Design 241(3), 968-977.

Bommer, J.J., F. Scherbaum, H. Bungum, F. Cotton, F. Sabetta and N.A. Abrahamson (2005). On the use of logic-trees for ground-motion prediction equations in seismic hazard assessment. *Bulletin of the Seismological Society of America* **95**(2), 377-389.

Bommer, J.J. and F. Scherbaum (2008). The use and misuse of logic trees in probabilistic seismic hazard analysis. *Earthquake Spectra* **24**(4), 997-1009.

Bommer, J.J., P.J. Stafford, J.E. Alarcón and S. Akkar (2007). The influence of magnitude range on empirical ground-motion prediction. *Bulletin of the Seismological Society of America* **97**(6), 2152-2170.

Bommer, J.J. and P.J. Stafford (2008). Seismic hazard and earthquake actions. *In*: Seismic Design of Buildings to Eurocode 8, A.Y. Elghazouli *ed.*, Taylor and Francis, 6-46.

Boore, D.M. (2003). Simulation of ground motion using the stochastic method. *Pure and Applied Geophysics* **160**, 635-676.

Boore, D.M. and G.M. Atkinson (2008). Ground-motion prediction equations for the average horizontal component of PGA, PGV, and 5%-damped PSA at spectral periods ranging from 0.01 to 10 s. *Earthquake Spectra* **24**(1), 99-138.

Boore, D.M. and J.J. Bommer (2005). Processing strong-motion accelerograms: needs, options and consequences. *Soil Dynamics & Earthquake Engineering* **25**(2), 93-115.

Boore, D.M., W.B. Joyner and T.E. Fumal (1997). Equations for estimating horizontal response spectra and peak acceleration from western North American earthquakes: a summary of recent work. *Seismological Research Letters* **68**, 128-153.

Bozorgnia, Y. and K.W. Campbell (2004a). The vertical-to-horizontal spectra ratio and tentative procedures for developing simplified V/H and vertical design spectra. *Journal of Earthquake Engineering* **8**, 175-207.

Bozorgnia, Y. and K.W. Campbell (2004b). Engineering characterization of ground motion. *In* Earthquake Engineering: From Engineering Seismology to Performance-Based Engineering, Y. Bozorgnia and V.V Bertero *eds.*, CRC Press.

Brune, J.N. (1999). Precarious rocks along the Mojave section of the San Andreas Fault, California: constraints on ground motion from great earthquakes. *Seismological Research Letters* **70**(1), 29-33.

Budnitz, R.J., G, Apostolakis, D.M. Boore, L.S. Cluff, K.J. Coppersmith, C.A. Cornell and P.A. Morris (1997). Recommendations for probabilistic seismic hazard analysis: guidance on uncertainty and the use of experts. *NUREG/CR-6372*, two volumes, US Nuclear Regulatory Commission, Washington, D.C.

Budnitz, R.J., Apostolakis, G., Boore, D.M., Cluff, L.S., Coppersmith, K.J., Cornell, C.A., Morris, P.A. (2006). Use of Technical Expert Panels: Applications to Probabilistic Seismic Hazard Analysis: *Risk Analysis* v. 18, no. 4, pp. 463 – 469, Society for Risk Analysis.

Cameron, W.I. and R.U. Green (2007). Damping correction factors for horizontal ground-motion response spectra. *Bulletin of the Seismological Society of America* **97**(3), 934-960.

Campbell, K.W. (1985). Strong motion attenuation relations: a ten-year perspective. *Earthquake Spectra* **1**(4), 759-804.

Campbell, K.W. (2003). Prediction of strong ground motion using the hybrid empirical method and its use in the development of ground motion (attenuation) relations in eastern North America. *Bulletin of the Seismological Society of America* **93**, 1012-1033.

Campbell, K.W. and Y. Bozorgnia (2008). NGA ground motion model for the geometric mean horizontal component of PGA, PGV, PGD and 5% damped linear elastic response spectra for periods ranging from 0.01 s to 10 s. *Earthquake Spectra* **24**(1), 139-171.

Chiou, B., R. Darragh, N. Gregor and W. Silva (2008). NGA project strong-motion database. *Earthquake Spectra* **24**(1), 23-44.

Chiou, B.S.-J. and R.R. Youngs (2008). An NGA model for the average horizontal component of peak ground motion and response spectra. *Earthquake Spectra* **24**(1), 173-215.

Chiou, B., R. Youngs, N. Abrahamson and K. Addo (2010). Ground-motion attenuation model for small-to-moderate shallow crustal earthquakes in California and its implication on regionalization of ground-motion prediction models. *Earthquake Spectra* **26**(4), 907-926.

Coats, D.W., and R.C. Murray (1984). Natural Phenomena Hazards Modeling Project: Seismic Hazard Models for Department of Energy Sites, *UCRL-53582*, Lawrence Livermore National Laboratory, University of California, Livermore, California.

Connor, C.B., N.A. Chapman and L.J. Connor eds. (2009). *Volcanic and Tectonic Hazard Assessment for Nuclear Facilities*. Cambridge University Press, 623 pp.

Coppersmith et al. (2009). Chapter 26, Lessons Learned—The Use of Formal Expert Assessment in Probabilistic Seismic and Volcanic Hazard Analysis," Coppersmith, K.J., Perman, R.C., Jenni, K.E., and Youngs, R.R., in *Volcanism, Tectonism, and Siting of Nuclear Facilities,* edited by C. Connor and L. Connor, Cambridge University Press, 593-611.

Coppersmith, K.J., J.J. Bommer, A.M. Kammerer, J. Ake. (2010). Implementation guidance for SSHAC Level 3 and 4 Processes *10th International Probabilistic Safety and Management Conference*, Seattle, Washington

Coppersmith, K.J., Perman, R.S., and Youngs, R.R. (2003). Earthquakes and Tectonics Expert Judgment Elicitation Project, EPRI-TR-102000, Electric Power Research Institute, Palo Alto CA, OSTI ID: 6635845

Coppersmith, K.J., R.R. Youngs, and C. Sprecher (2009). Methodology and main results of seismic source characterization for the PEGASOS project, Switzerland: *Swiss Journal of Geosciences*, v. 102. 91-105.

Coppersmith, K.J., and Youngs, R.R. (1990), Probabilistic Seismic Hazard Analysis Using Expert Opinion: An Example from the Pacific Northwest, *in* Krinitsky, E.L., and Slemmons, D.B., *Neotectonics in Earthquake Evaluation, Geological Society of America Reviews in Engineering Geology,* vol. 8, Boulder, Colorado.

Coppersmith, K.J. and R.R. Youngs (1986). Capturing uncertainty in probabilistic seismic hazard assessments within intraplate tectonic environments. *Proceedings of the Third US National Conference on Earthquake Engineering,* vol. 1, 301-312.

Cornell, C.A. (1968). Engineering seismic risk analysis. *Bulletin of the Seismological Society of America* **58**(5), 1583-1606. Erratum: **59**(4), 1733.

Cornell, C.A. (1971) Probabilistic analysis of damage to structures under seismic loads. In *Dynamic Waves in Civil Engineering, eds.* D.A. Howells, I.P. Haigh and C. Taylor, John Wiley and Sons, 473-488.

Cornell, C.A. and H.A. Merz (1975) Seismic risk analysis of Boston. *Journal of the Structural Division ASCE* **101**(ST10), 2027-2043.

Cotton, F., F. Scherbaum, J.J. Bommer and H. Bungum (2006). Criteria for selecting and adjusting ground-motion models for specific target applications: applications to Central Europe and rock sites. *Journal of Seismology* **10**(2), 137-156

Crowley, H. and J.J. Bommer (2006). Modeling seismic hazard in earthquake loss models with spatially distributed exposure. *Bulletin of Earthquake Engineering* **4**(3), 249-275.

CRWMS M&O (1996). Probabilistic Volcanic Hazard Analysis for Yucca Mountain, Nevada. BA0000000-01717-2200-00082 REV 0. Las Vegas, Nevada: CRWMS M&O.

CRWMS M&O (1997). *Unsaturated Zone Flow Model Expert Elicitation Project.* Las Vegas, Nevada: CRWMS M&O.

CRWMS M&O (1998a). *Near-Field/Altered Zone Coupled Effects Expert Elicitation Project.* Las Vegas, Nevada: CRWMS M&O.

CRWMS M&O (1998b). *Probabilistic Seismic Hazard Analyses for Fault Displacement and Vibratory Ground Motion at Yucca Mountain, Nevada.* Milestone SP32IM3, September 23, 1998. Three volumes. Las Vegas, Nevada: CRWMS M&O.

CRWMS M&O (1998c). *Saturated Zone Flow and Transport Expert Elicitation Project.* Deliverable SL5X4AM3. Las Vegas, Nevada: CRWMS M&O.

CRWMS M&O (1998d). *Waste Form Degradation and Radionuclide Mobilization Expert Elicitation Project.* Las Vegas, Nevada: CRWMS M&O.

CRWMS M&O (1998e). *Waste Package Degradation Expert Elicitation Project.* Rev. 1. Las Vegas, Nevada: CRWMS M&O.

Dalkey, N.C. (1967), *Delphi,* The RAND Corporation, Report P 3704, Santa Monica, CA.

Dalkey, N.C. (1969), *The Delphi Method: An Experimental Study of Group Opinion,* The RAND Corporation, Report RM-5888-PR, Santa Monica, CA.

Dalkey, N.C., and Helmer, O. (1963), An Experimental Application of the DEPHI Method to the Use of Experts, *Management Science,* 9, pp. 458-467.

Delavaud, E., F. Scherbaum, N. Kuehn and C. Riggelsen (2009). Information-theoretic selection of ground-motion prediction equations for seismic hazard analysis: an applicability study using Californian data. *Bulletin of the Seismological Society of America* **99**(6), 3248-3263.

Department of Energy (DOE) (2008). Yucca Mountain Repository License Application, DOE/RW-0573, Update No. 1, Docket No. 63–001.

DeWispelare, A.R., L.T. Herren, M.P. Miklas, and R.T. Clemen (1993). Expert Elicitation of Future Climate in the Yucca Mountain Vicinity. Report NRC-02-88-005, Center for Nuclear Waste Regulatory Analyses, San Antonio, Texas.

DOE (U.S. Department of Energy) (1998). *Viability Assessment of a Repository at Yucca Mountain.* DOE/RW-0508. Overview and five volumes. Washington, D.C.: U.S. Department of Energy, Office of Civilian Radioactive Waste Management.

DOE O 420.1B. Reviewed December 2007. *Facility Safety.* Department of Energy Order Washington, DC.

Douglas, J. (2003). Earthquake ground motion estimation using strong-motion records: a review of equations for the estimation of peak ground acceleration and response spectral ordinates. *Earth-Science Reviews* **61**, 43-104.

Douglas, J. (2011). Ground-motion prediction equations 1964-2010. Report BRGM/RP-59356-FR, BGRM, Orléans, France.

Elms, D.G. (2004). Structural safety – issues and progress. *Progress in Structural Engineering and Materials* **6**, 116-126.

EPRI (1989). Engineering characterization of small-magnitude earthquakes. *EPRI Report NP-6389,* Electric Power Research Institute, Palo Alto, California.

EPRI-SOG (1988). Seismic Hazard Methodology for the Central and Eastern United States, EPRI NP-4726A, Revision 1, Volumes 1-11, Electric Power Research Institute, Palo Alto, California.

EPRI (1989). Probabilistic Seismic Hazard Evaluations at Nuclear Plant Sites in the Central and Eastern United States: Resolution of the Charleston Earthquake Issue, EPRI NP-6395-D. Palo Alto, CA: Electric Power Research Institute.

EPRI (2006). Use of CAV in determining effects of small magnitude earthquakes on seismic hazard analysis. *EPRI Report TR-1014099,* Electric Power Research Institute, Palo Alto, California.

EPRI (2008) Project Plan: Central and Eastern United States Seismic Source Characterization for Nuclear Facilities, Technical Update1016756.

EPRI (2004). CEUS Ground Motion Project Final Report, EPRI Report 1009684, Electric Power Research Institute, Palo Alto, CA.

EPRI (2006). Program on Technology Innovation: Truncation of the Lognormal Distribution and Value of the Standard Deviation for Ground Motion Models in the central and Eastern United States, EPRI Report 1013105, Technical Update, Electric Power Research Institute, Palo Alto, CA.

Esteva, L. (1969) Seismicity prediction: a Bayesian approach. *Proceedings of the Fourth World Conference on Earthquake Engineering*, Santiago de Chile, vol. 1, A-1, 172-185.

Esteva, L. (1970) Seismic risk and seismic design decisions. In *Seismic Design for Nuclear Power Plants, ed.* R.J. Hansen, MIT Press, Cambridge, Massachusetts, 142-182.

Frankel, A. (1995). Mapping seismic hazard in the Central and Eastern United States. *Seismological Research Letters* **66**, 8-21.

Fujiwara, H., N. Morikawa, Y. Ishikawa, T. Okumura, J. Miyakoshi, N. Nojima and Y. Fukushima (2009). Statistical comparison of national probabilistic seismic hazard maps and frequency of recorded JMA seismic intensities from the K-NET strong-motion observation network in Japan during 1997-2006. *Seismological Research Letters* **80**(3), 458-464.

Gneiting, T. and A. Raferty (2007). Strictly proper scoring rules, prediction, and estimation. *Journal of the American Statistical Association* **102**, 359-378.

Hancock, J. and J.J. Bommer (2005). The effective number of cycles of earthquake ground motion. *Earthquake Engineering & Structural Dynamics* **34**(6), 637-664.

Hanks, T.C. and W.H. Bakun (2002). A bilinear source-scaling model for M-log$A$ observations of continental earthquakes. *Bulletin of the Seismological Society of America* **92**(5), 1841-1846.

Hanks, T.C. and H. Kanamori (1979). A moment magnitude scale. *Journal of Geophysical Research* **77**(23), 4393-4405.

Hanks, T.C., N.A. Abrahamson, D.M. Boore, K.J. Coppersmith and N.E. Knepprath (2009). Implementation of the SSHAC Guidelines for Level 3 and 4 PSHAs – Experience gained from actual applications. *Open-File Report 2009-1093*, US Geological Survey, Reston VA.

Hofer, E. (1996). When to separate uncertainties and when not to separate. *Reliability Engineering and System Safety* **54**, 113-118.

Hora, S.C. (1993). Acquisition of Expert Judgment: Examples from Risk Assessment, Journal of Energy Engineering, 118:136-148 .

Idriss, I. M. (1978). Characteristics of earthquake ground motions. *Proceedings of the ASCE Specialty Conference on Earthquake Engineering and Soil Dynamics-Pasedena, California*, **V III**, 1154-1261, American Society of Civil Engineers, New York.

Joyner, W.B. and D.M. Boore (1988). Measurement, characterization, and prediction of strong ground motion. *Proceedings of ASCE/GT Div. Specialty Conference on Soil Dynamics & Earthquake Engineering II*, Park City, Utah. 43-102.

Kaplan, S. (1992), "Expert Information Versus Expert Opinions: Another Approach to the Problem of Eliciting/Combining Using Expert Knowledge in PRA," *Journal of Reliability Engineering and System Safety* **35**, 61-72.

Keefer, D.K. (1984). Landslides caused by earthquakes. *Bulletin of the Geological Society of America* **95**, 406-421.

Keefer, D.L. and S.E. Bodily (1983). Three-point approximations for continuous random variables. *Management Science* **29**(5), 595-609.

Keeney, R.L. (1996). Value-Focused Thinking: A Path to Creative Decisionmaking. *Harvard University Press*, Cambridge, Ma, 432p.

Kennedy, R.P. (1999). Overview of methods for seismic PRA and margin analysis including recent innovations. *Proceedings of the OECD-NEA Workshop on Seismic Risk*, August 10-12, Tokyo, Japan.

Kennedy, R.P. (2007). Performance-goal based (risk informed) approach for establishing the SSE site specific spectrum for future nuclear power plants. *Transactions of 19th International Conference on Structural Mechanics in Reactor Technology (SMiRT 19)*, August 12-17, Toronto, Canada.

Kerr, R. A. (1996). Risk Assessment—A new way to ask the experts: Rating radioactive waste risks: Science**274** (5289), 913-914.

Klügel, J.-U. (2005). Problems in the application of the SSHAC probability method for assessing earthquake hazards at Swiss nuclear power plants. Engineering Geology 78, 285–307.

Kotra, J.P., Lee, M.P., Eisenberg, N.A., and DeWispelare, A.R (1996). Branch technical position on the use of expert elicitation in the High-Level Radioactive Waste Program: Washington, D.C., U.S. Nuclear Regulatory Commission, NUREG-1563.

Kulkarni, R.B., R.R. Youngs and K.J. Coppersmith (1984). Assessment of confidence intervals for results of seismic hazard analysis. *Proceedings of the Eighth World Conference on Earthquake Engineering*, San Francisco vol. 1, 263-270.

Leonard, M. (2010). Earthquake fault scaling: self-consistent relating of rupture length, width, average displacement, and moment release. *Bulletin of the Seismological Society of America* **100**(5A),1971-1988.

Matthews, M.V., W.L. Ellsworth and P.A. Reasenberg (2002). A Brownian model for recurrent earthquakes. *Bulletin of the Seismological Society of America* **92**(6), 2233-2250.

McGuire, R.K. (1976) FORTRAN computer program for seismic risk analysis. *U.S. Geological Survey. Open-File Report 76-67.*

McGuire, R.K. (1977). Seismic design and mapping procedures using hazard analysis based directly on oscillator response. *Earthquake Engineering & Structural Dynamics* **5**, 211-234.

McGuire, R.K. (1993). Computations of seismic hazard. *Annali di Geofisica* **36**(3-4), 181-200.

McGuire, R.K. (1995). Probabilistic seismic hazard analysis and design earthquakes: closing the loop. *Bulletin of the Seismological Society of America* **85**(5), 1275-1284.

McGuire, R.K. (2004). Seismic hazard and risk analysis. *EERI Monograph MNO-10*, Earthquake Engineering Research Institute, Oakland, California.

McGuire, R.K. (2008). Probabilistic seismic hazard analysis: early history. *Earthquake Engineering & Structural Dynamics* **37**, 329-338.

McGuire, R. (2009). Issues in Probabilistic Seismic Hazard Analysis for Nuclear Facilities in the US, *Nuclear Engineering and Technology*, v.41, no.10, p. 1235-1242.

McGuire, R.K., W.J. Silva and C.J. Costantino (2001). Technical basis for revision of regulatory guidance on design ground motions: hazard- and risk-consistent ground motion spectra guidelines. *NUREG/CR-6728*, US Nuclear Regulatory Commission, Washington D.C.

Merz, H.A. and C.A. Cornell (1973) Seismic risk based on a quadratic magnitude-frequency law. *Bulletin of the Seismological Society of America* **63**(6), 1999-2006.

Meyer, M.A. and J.M. Booker, (1990) Eliciting and Analyzing Expert Judgment: A Practical Guide, U.S. Nuclear Regulatory Commission, NUREG/CR-5424. [Prepared by the Los Alamos National Laboratory.]

Miyazawa, M. and J. Mori (2009). Test of seismic hazard map from 500 years of recorded intensity data in Japan. *Bulletin of the Seismological Society of America* **99**(6), 3140-3149.

Morgan, M.G. and D.W. Keith (1995). Subjective judgments by climate experts. *Environmental Science and Technology* **29**, A468-A476.

Morgan, M.G., and M. Henrion (1990). Uncertainty: A Guide to Dealing with Uncertainty in Quantitative Risk and Policy Analysis, Cambridge, Massachusetts, Cambridge University Press.

Musson, R.M.W. (2004). Objective validation of seismic hazard source models. *Proceedings 13th World Conference on Earthquake Engineering*, August 1-6, Vancouver, Canada, paper no. 2492.

Musson, R.M.W., Toro, G.R., Coppersmith, K.J., Bommer, J.J., Deichmann, N., Bungum, H., Cotton, F., Scherbaum, F., Slejko, D., Abrahamson, N.A. (2005). Evaluating hazard results for Switzerland and how not to do it: A discussion of problems in the application of the SSHAC probability method for assessing earthquake hazards at Swiss nuclear power plants by J.-U.Klügel., Eng. Geol. Vol. 78, pp. 285–307.

NAGRA (2004). Probabilistic Seismic Hazard Analysis for Swiss Nuclear Power Plant Sites (PEGASOS Project): Report to Swissnuclear prepared by Nationale Genossenschaft für die Lagerung radioaktiver Abfälle, Wettingen, 358 p.

O'Hagan, A., C.E. Buck, A. Daneshkhah, J.R. Eiser, P.H. Garthwaite, D.J. Jenkinson, J.E. Oakley and T. Rakow (2006). *Uncertain Judgments: Eliciting Experts' Probabilities*. John Wiley & Sons, Chichester, UK, 321 pp.

Okrent, D (1975). A Survey of Expert Opinion on Low Probability Earthquakes. *Annals of Nuclear Energy, Vol. 2,* Pergamon Press, Ireland, pp. 601-614.

Ordaz, M. and C. Reyes (1999). Earthquake hazard in Mexico City: observations versus computations. *Bulletin of the Seismological Society of America* **89**(5), 1379-1383.

Oreskes, N., K. Shrader-Frechette and K. Belitz (1994). Verification, validation, and confirmation of numerical models in the Earth sciences. *Science* **263** (5147), 641-646.

Pancha, A., J.G. Anderson and C. Kreemer (2006). Comparison of seismic and geodetic scalar moment rates across the Basin and Range provinces. *Bulletin of the Seismological Society of America* **96**(1), 11-32.

Parsons, T. and E.L. Giest (2009). Is there a basis for preferring characteristic earthquakes over a Gutenberg-Richter distribution in probabilistic earthquake forecasting? *Bulletin of the Seismological Society of America* **99**(3), 2012-2019.

Petersen, M., T. Cao, T. Dawson, A. Frankel, C. Wills and D. Schwartz (2004). Evaluating fault rupture hazard for strike-slip earthquakes. *Proceedings of Geo-Trans 2004*, July 27-31, Los Angeles, *ASCE Geotechnical Special Publication no. 126*, 787-796.

Petersen, M.D., C.H. Cramer, M.S. Reichle, A.D. Frankel and T.C. Hanks (2000). Discrepancy between earthquake rates implied by historic earthquakes and a consensus geologic source model for California. *Bulletin of the Seismological Society of America* **90**(5), 1117-1132.

Petersen, M.D., A.D. Frankel, S.C. Harmsen, C.S. Mueller, K.M. Haller, R.L. Wheeler, R.L. Wesson, Y. Zeng, O.S. Boyd, D.M. Perkins, N. Luco, E.H. Field, C.J. Wills and K.S. Rukstales (2008). Documentation for the 2008 update of the United States national seismic hazard maps. *USGS Open-File Report 2008-1128, US Geological Survey*, Reston, Virginia.

PG&E (2010). Methodology for probabilistic tsunami hazard analysis: trial application for the Diablo Canyon power plant site. *Draft Report*, February, Pacific Gas and Electricity, San Francisco, California.

Power, M., B. Chiou, N. Abrahamson, Y. Bozorgnia, T, Shantz and C. Roblee (2008). An overview of the NGA project. *Earthquake Spectra* **24**(1), 3-21.

Reed, J.W. and R.P. Kennedy (1988). A criterion for determining exceedance of the operating basis earthquake. *EPRI Report NP-5935*, Electric Power Research Institute, Palo Alto, California.

Reiter, L. (1990). Earthquake Hazard Analysis: Issues and Insights. Columbia University Press, New York.

Reiter, L. and R. E. Jackson, (1983), Seismic Hazard Review for the Systematic Evaluation Program: A Use of Probability in Decision Making. U.S. Nuclear Regulatory Commission, NUREG-0967, 55 p.

Rikitake, T. and I. Aida (1988). Tsunami hazard probability in Japan. *Bulletin of the Seismological Society of America* **78**(3), 1268-1278.

Rodríguez, C.E., J.J. Bommer and R.J. Chandler (1999). Earthquake-induced landslides 1980-1997. *Soil Dynamics and Earthquake Engineering* **18**(5), 325-346.

Sabetta, F., A. Lucantoni, H. Bungum and J.J. Bommer (2005). Sensitivity of PSHA results to ground motion prediction relations and logic-tree weights. *Soil Dynamics & Earthquake Engineering* **25**(4), 317-329.

Savy, J.B., A. C. Boissonnade, R. W. Mensing, and C. M. Short (1993). "Eastern Seismic Hazard Characterization Update," UCRL-ID-115111, Lawrence Livermore National Laboratory, June 1993 (Also published as U.S. NRC [U.S. Nuclear Regulatory Commission], 1993, Revised Livermore seismic hazard estimates for 69 nuclear power plant sites east of the Rocky Mountains: Washington, D.C., U.S. Nuclear Regulatory Commission Report, NUREG–1488).

Savy, J.B., Foxall, W., Abrahamson, N., Bernreuter, D., Zurflueh, E., and Pullani, S. (2002). *Guidance for Performing Probabilistic Seismic Hazard Analysis for a Nuclear Plant Site: Example Application to the Southeastern United States.* NUREG/CR-6607. Washington, D.C.: U.S. Nuclear Regulatory Commission, Office of Nuclear Regulatory Research.

Scherbaum, F., J.J. Bommer, H. Bungum, F. Cotton and N.A. Abrahamson (2005). Composite ground-motion models and logic-trees: methodology, sensitivities and uncertainties. *Bulletin of the Seismological Society of America* **95**(5), 1575-1593.

Scherbaum, F., F. Cotton and P. Smit (2004b). On the use of response spectral reference data for the selection and ranking of ground motion models for seismic hazard analysis in regions of low to moderate seismicity. *Bulletin of the Seismological Society of America* **94**, 2164-2185.

Scherbaum, F., F. Cotton and H. Staedtke (2006). The estimation of minimum-misfit stochastic models from empirical prediction equations. *Bulletin of the Seismological Society of America* **96**(2), 427-445.

Scherbaum, F., E. Delavaud and C. Riggelsen (2009). Model selection in seismic hazard analysis: an information-theoretic perspective. *Bulletin of the Seismological Society of America* **99**(6), 3234-3247.

Scherbaum, F., J. Schmedes and F. Cotton (2004a). On the conversion of source-to-site distance measures for extended earthquake source models. *Bulletin of the Seismological Society of America* **94**, 1053-1069.

Schlaifer, R.O. (1959), *Probability and Statistics for Business decisions: An Introduction to Managerial Economics under Uncertainty,* McGraw-Hill, New York.

Schorlemmer, D., M.C. Gerstenberger, S. Wiemer, D.D. Jackson and D.A. Rhoades (2007). Earthquake likelihood model testing. *Seismological Research Letters* **78**(1), 17-29.

Schwartz, D.P. and K.J. Coppersmith (1984). Fault behavior and characteristic earthquakes: examples from the Wasatch and San Andreas faults. *Journal of Geophysical Research* **89**, 5681-5698.

SNL (2008). Probabilistic volcanic hazard analysis update (PVHA-U) for Yucca Mountain, Nevada. *Report TDR-MGR-PO-000001 (Rev 00)*, Sandia National Laboratory, Las Vegas, Nevada.

Sobel, P. (1994). Revised Livermore Seismic Hazard Estimates for Sixty Nine Nuclear Power Plant Sites East of the Rocky Mountains. *NUREG-1488*, US Nuclear Regulatory Commission, Washington, D.C.

Spetzler, C.S, and Staël von Holstein , C.A.S (1975), Probability Encoding in Decision Analysis, *Management Science*, 22(3), pp. 340-358.

Stafford, P.J., F.O. Strasser and J.J. Bommer (2008b). An evaluation of the applicability of the NGA models to ground-motion prediction in the Euro-Mediterranean region. *Bulletin of Earthquake Engineering* 6(2), 149-177.

Stafford, P., R. Mendis and J.J. Bommer (2008a). Dependence of damping correction factors for response spectra on duration and number of cycles. *ASCE Journal of Structural Engineering* 134(8), 1364-1373.

Stein, R.S. and T.C. Hanks (1998). M 6 earthquakes in southern California during the 20th century: no evidence for a seismicity or moment deficit. *Bulletin of the Seismological Society of America* 88, 635-652.

Stepp, J.C., I. Wong, J. Whitney, R. Quittemeyer, N. Abrahamson, G. Toro, R. Youngs, K. Coppersmith, J. Savy, T. Sullivan and Yucca Mountain PSHA Project Members (2001). Probabilistic seismic hazard analyses for ground motions and fault displacements at Yucca Mountain, Nevada. *Earthquake Spectra* 17(1), 113-151.

Stirling, M.W. and R. Anooshehpoor (2006). Constraints on probabilistic seismic-hazard models from unstable landform features in New Zealand. *Bulletin of the Seismological Society of America* 96(2), 404-414.

Stirling, M. and M. Petersen (2006). Comparison of the historical record of earthquake hazard with seismic-hazard models for New Zealand and the continental United States. *Bulletin of the Seismological Society of America* 96(4), 1978-1994.

Stirling, M. and M. Gerstenberger (2010). Ground motion-based testing of seismic hazard models in New Zealand. *Bulletin of the Seismological Society of America* 100(4), 1407-1414.

Strasser, F.O., J.J. Bommer and N.A. Abrahamson (2008). Truncation of the distribution of ground-motion residuals. *Journal of Seismology* 12(1), 79-105.

Strasser, F.O., N.A. Abrahamson and J.J. Bommer (2009). Sigma: issues, insights, and challenges. *Seismological Research Letters* 80(1), 40-56.

Strasser, F.O., M.C. Arango and J.J. Bommer (2010). Scaling of the source dimensions of interface and intraslab subduction-zone earthquakes with moment magnitude. *Seismological Research Letters* 81(6), 941-950.

Thenhaus, P.C. and K.W. Campbell (2003). Seismic hazard analysis. *In*: Earthquake Engineering Handbook, W.F. Chen and C. Scawthorn *eds.*, CRC Press, Boca Raton, Florida, 8.1-8.50.

Thomas, P., I. Wong and N. Abrahamson (2010). Verification of probabilistic seismic hazard analysis computer programs. *PEER Report 2010/106*, Pacific Earthquake Engineering Research Center, UB Berkeley, California.

Trauth, K.M., R.P. Rechard, and S.C. Hora (1991). "Expert Judgment as Input to Waste Isolation Pilot Plant Performance-Assessment Calculations: Probability Distributions of Significant System Parameters," Mixed Waste: Proceedings of the First International Symposium, Baltimore, MD, August 26-29, 1991. Eds. A.A. Moghissi and G. A. Benda. SAND91-0625C. Baltimore, MD: Environmental Health and Safety, University of Maryland. 4.3.1 - 4.3.9.

Toro, G.R., N.A. Abrahamson and J.F. Schneider (1997). Models of strong ground motions from earthquakes in Central and Eastern North America: best estimates and uncertainties. *Seismological Research Letters* **68**(1), 41-57.

US Environmental Protection Agency (1985) "Environmental Standards for the Management and Disposal of Spent Nuclear Fuel, High-Level and Transuranic Radioactive Wastes [Final Rule], Federal Register, Vol. 50, No. 182, pp. 38066–38089, September 19, 1985.

USNRC (1975). Reactor Safety Study, An Assessment of Accident Risks in U.S. Commercial Nuclear Power Plants, WASH-1400 (NUREG-75.014), U.S. Nuclear Regulatory Commission, Washington D.C., 228p. [commonly referred to as the Rasmussen Report]

USNRC (1991a). Severe Accident Risks: An Assessment for Five U.S. Nuclear Power Plants — Final Summary Report (NUREG-1150, Volume 1), Division of Systems Research Office of Nuclear Regulatory Research U.S. Nuclear Regulatory Commission Washington, DC 20555

USNRC (1991b), Staff's Approach for Dealing with Uncertainties in Implementing the EPA HLW Standards, Commission Paper (Policy/Information) SECY-91-242, August 6, 1991.

USNRC (1997). Regulatory Guide 1.165, "Identification and Characterization of Seismic Sources and Determination of Safe Shutdown Earthquake Ground Motion," U.S. Nuclear Regulatory Commission, Washington, DC.

USNRC (2007). A Performance-Based Approach to Define the Site-Specific Earthquake Ground Motion. *Regulatory Guide 1.208*, US Nuclear Regulatory Commission, Washington D.C.

USNRC (2010), Resolution of Generic Safety Issues: Issue 194: Implications of Updated Probabilistic Seismic Hazard Estimates (NUREG-0933, Main Report with Supplements 1–32), US Nuclear Regulatory Commission, Washington D.C.

Utsu. T. (2002). Relationships between magnitude scales. *In*: International Handbook of Earthquake and Engineering Seismology, *eds*. W.H.K. Lee, H. Kanamori, P.C. Jennings and C. Kisslinger, Part A, 733-746, Academic Press.

Veneziano, D., A. Agarwal and E. Karaca (2009). Decision making with epistemic uncertainty under safety constraints: An application to seismic design. *Probabilistic Engineering Mechanics* **24**, 426-437.

Ward, S.N. (1995). Area-based tests of long-term seismic hazard predictions. *Bulletin of the Seismological Society of America* **85**(5), 1285-1298.

Wells, D.L. and K.J. Coppersmith (1994). New empirical relationships among magnitude, rupture length, rupture width, rupture area, and surface displacement. *Bulletin of the Seismological Society of America* **84**(4), 974-1002.

Wesnousky, S.G. (1994). The Gutenberg-Richter or characteristic earthquake distribution, which is it? *Bulletin of the Seismological Society of America* **84**(6), 1940-1959.

Woo, G. (1996). Kernel estimation methods for seismic hazard area source modeling. *Bulletin of the Seismological Society of America* **86**, 353-362.

Youngs, R.R. and K.J. Coppersmith (1985). Implications of fault slip rates and earthquake recurrence models to probabilistic seismic hazard estimates. *Bulletin of the Seismological Society of America* **75**, 939-964.

Youngs, R.R., W. Arabasz, R.E. Anderson, A.R. Ramelli, J.P. Ake, D.B. Slemmons, J.P. McCalpin, D.I. Doser, C.J. Fridrich, F.H. Swan, A.M. Rogers. J.C. Yount, L.W. Anderson. K.D. Smith, R.L. Bruhn, P.L.K. Knuepfer, R.B. Smith, C.M. dePolo, D.W. O'Leary, K.J. Coppersmith, S.K. Pezzopane, D.P. Schwartz, J.W. Whitney, S.S. Olig and G.R. Toro (2003). A methodology for probabilistic fault displacement hazard analysis (PFDHA). *Earthquake Spectra* **19**(1), 191-219.

APPENDIX A: Executive Summary of NUREG/CR-6372

Probabilistic seismic hazard analysis (PSHA) is a methodology that estimates the likelihood that various levels of earthquake-caused ground motions will be exceeded at a given location in a given future time period. The results of such an analysis are expressed as estimated probabilities per year or estimated annual frequencies. The objective of this project has been to provide methodological guidance on how to perform a PSHA. The project, co-sponsored by the U.S. Nuclear Regulatory Commission, the U.S. Department of Energy, and the Electric Power Research Institute, has been carried out by a seven member Senior Seismic Hazard Analysis Committee (SSHAC), supported by a large number of other experts working under the Committee's guidance, who are named in the following "Acknowledgments" section.

The members of the Senior Seismic Hazard Analysis Committee (SSHAC) are:

Dr. Robert J. Budnitz (Chairman)	President Future Resources Associates, Inc.
Professor George Apostolakis	Massachusetts Institute of Technology Previously at University of California, Los Angeles
Dr. David M. Boore	Seismologist U.S. Geological Survey
Dr. Lloyd S. Cluff	Manager, Geosciences Department Pacific Gas & Electric Company
Dr. Kevin J. Coppersmith	Vice President Geomatrix
Dr. C. Allin Cornell	C. A. Cornell Company
Dr. Peter A. Morris	Applied Decision Analysis, Inc.

The scope of the SSHAC guidance is intended to cover both site-specific and regional applications of PSHA (more broadly, applications in both low-seismicity and high-seismicity regions) in both the eastern U.S. and western U.S. Although the sponsors' primary objective is guidance for applications at nuclear power plants and other critical facilities, the methodological guidance applies in whole or in part, on a case-by-case basis, to a broad range of applications.

The SSHAC guidance involves both technical guidance and procedural guidance, with a strong emphasis on the latter for reasons explained below. Therefore, the audience for the report includes not only analysts who will implement the methodology and earth scientists whose expertise will support the analysts, but also PSHA project sponsors—those decision-makers in organizations such as private firms or government agencies who have a need for PSHA information and are in a position to sponsor a PSHA study.

Note that our guidance is not intended to be "the only" or "the standard" methodology for PSHA to the exclusion of other approaches; there are other valid ways to perform a PSHA study. Likewise, our formulation should not be viewed as an attempt to "standardize" PSHA in the sense of freezing the science and technology that underlies a competent PSHA, thereby stifling innovation. Rather, our guidance is intended to represent SSHAC's opinion on the best current thinking on performing a valid PSHA.

The most important and fundamental fact that must be understood about a PSHA is that the objective of estimating annual frequencies of exceedance of earthquake-caused ground motions can be attained only with significant uncertainty. Despite much recent research, major gaps exist in our understanding of the mechanisms that cause earthquakes and of the processes that govern how an earthquake's energy propagates from its origin beneath the earth's surface to various points near and far on the surface. The limited information that does exist can be—and often is—legitimately interpreted quite differently by different experts, and these differences of interpretation translate into important uncertainties in the numerical results from a PSHA.

The existence of these differences of interpretation translates into an operational challenge for the PSHA analyst who is faced with (1) how to use these different interpretations properly, and (2) how to incorporate the diversity of expert judgments into an analytical result that appropriately captures the current state-of-knowledge of the expert community, including its uncertainty.

The SSHAC studied a large number of past PSHAs, including two landmark studies from the late 1980s known as the "Lawrence Livermore (LLNL)" study and the "Electric Power Research Institute (EPRI)" study, both of which broke important new methodological ground in attempting to characterize earthquake-caused ground motion in the broad region of the U.S. east of the Rocky Mountains. Most important, the mean seismic hazard curves presented in the reports for most sites in the eastern U.S. differed significantly. However, the median hazard results did not differ by nearly as much. We now understand that differences in both the inputs and the procedures by which the two studies dealt with the inputs were among the key reasons for the differences in the mean curves. At the time this was not understood, and the differences between the mean curves caused not only considerable consternation, but launched several efforts to understand what might underlie the differences and attempts to update the older work.

Ultimately, the inability to understand all of the differences between the LLNL and EPRI hazard results and the concomitant need for an improved methodology going beyond the late-1980s state-of-the-art-led directly to the formation of the SSHAC to perform this project. However, although the Committee studied both the LLNL and EPRI projects carefully to obtain methodological insights (both positive and negative), it did not undertake a forensic-type review to identify past "errors." Rather, it attempted to draw more broadly upon the entire body of PSHA literature and experience, including of course the LLNL and EPRI projects along with many others, to formulate the guidance herein.

In the course of our review, we concluded that many of the major potential pitfalls in executing a successful PSHA are procedural rather than technical in character. One of the most difficult challenges for the PSHA analyst is properly representing the wide diversity of expert judgments about the technical issues in PSHA in an acceptable analytical result, including addressing the large uncertainties. This conclusion, in turn, explains our heavy emphasis on procedural guidance.

This also explains why we believe that how a PSHA is structured is as critical to its success as the technical aspects—perhaps more critical because the procedural pitfalls can sometimes be harder to avoid and harder to uncover in an independent review than the pitfalls in the technical aspects. Finally, this also explains why one of the key audiences for this report is the project sponsor, who needs to understand the procedural/structural aspects in order to initiate and support the desired PSHA project appropriately.

This Executive Summary will conclude with a brief overview of what the SSHAC believes are its most important findings, conclusions, and recommendations in the procedural area. Because we recognize that several very important pieces of technical guidance concerning the earth-sciences aspects of PSHA will not be discussed in this Executive Summary, the SSHAC requests that readers turn to the full report to review the technical guidance. The key procedural points follow:

1. SSHAC identifies and describes several different *roles for experts* based on its conclusion that confusion about the various roles is a common source of difficulty in executing the aspect of PSHA involving the use of experts. The roles for which SSHAC provides the most extensive guidance include the expert as *proponent* of a specific technical position, as an *evaluator* of the various positions in the technical community, and as a *technical integrator* (see the next paragraph).

2. SSHAC identifies four different types of consensus, and then concludes that one key source of difficulty is failure to recognize that 1) there is not likely to be "consensus" (as the word is commonly understood) among the various experts and 2) no single interpretation concerning a complex earth sciences issue is the "correct" one. Rather, SSHAC believes that the following should be sought in a properly executed PSHA project for a given difficult technical issue: (1) a representation of the legitimate range of technically supportable interpretations among the entire informed technical community, and (2) the relative importance or credibility that should be given to the differing hypotheses across that range. As SSHAC has framed the methodology, this information is what the PSHA practitioner is charged to seek out, and seeking it out and evaluating it is what SSHAC defines *as technical integration.*

3. SSHAC identifies a hierarchy of complexity for technical issues, consisting of four *levels* (representing increasing levels of participation by technical experts in the development of the desired results), and then concentrates much of its guidance on the most complex level (level 4) in which a panel of experts is formally constituted and the panel's interpretations of the technical information relevant to the issues are formally elicited. To deal with such complex issues, SSHAC defines an entity that it calls the Technical Facilitator/Integrator (TFI), which is differentiated from a similar entity for dealing with issues at the other three less-complex levels, which SSHAC calls the Technical Integrator (TI). Much of SSHAC's procedural guidance involves how the TI and TFI functions should be structured and implemented. (Both the TI and TFI are envisioned as roles that may be filled by one person or, in the TFI case, perhaps by a small team).

4. The role of *technical integration* is common to the TI and TFI roles. What is special about the TFI role, in SSHAC's formulation, is the *facilitation* aspect, when an issue is judged to be complex enough that the views of a panel of several experts must be elicited. SSHAC's guidance dwells on that aspect extensively, in part because SSHAC believes that this is where some of the most difficult procedural pitfalls are encountered. In fact, the main report identifies a number of problems that have arisen in past PSHAs and discusses how the TFI function explicitly overcomes each of them.

5. For most technical issues that arise in a typical PSHA, the issue's complexity does not warrant a panel of experts and hence the establishment of a TFI role. Technical integration for these issues can be accomplished—indeed, is usually best accomplished—by a TI. In fact, SSHAC has structured its recommended methodology so that even the most complex issues *can* be dealt with using the less expensive TI mode, although with some sacrifice in the confidence obtained in the results on both the technical and the procedural sides.

6. One special element of the TFI process is SSHAC's guidance on sequentially using the panel of experts in different roles. Heavy emphasis is placed on assuring constructive give-and-take interactions among the panelists throughout the process. Each expert is first asked, based on his/her own knowledge (yet cognizant of the views of others as explored through the information-exchange process), to act as an *evaluator;* that is, to evaluate the range of technically legitimate viewpoints concerning the issue at hand. Then, each expert is asked to play the role of *technical integrator,* providing advice to the TFI on the appropriate representation of the composite position of the community as a whole. Contrasting the classical role of experts on a panel acting as individuals and providing inputs to a separate aggregation process, the TFI approach views the panel as a team, with the TFI as the team leader, working together to arrive at (i) a composite representation of the knowledge of the group, and then (ii) a composite representation of the knowledge of the technical community at large. (Neither of these representations necessarily reflects panel consensus—they may or may not, and their validity does *not* depend on whether a panel consensus is reached.) The SSHAC guidance to the TFI emphasizes that a variety of techniques are available for achieving this composite representation. SSHAC recommends a blending of behavioral or judgmental methods with mathematical methods, and in the body of the report several techniques along these lines are described in detail. A key objective for the TFI is to develop an aggregate result that can be endorsed by the expert panel both technically and in terms of the process used.

7. The TFI's integrator role should be viewed not as that of a "super-expert" who has the final say on the weighting of the relative merits of either specific technical interpretations or the various experts' interpretations of them; rather, the TFI role should be seen as charged with characterizing both the commonality and the diversity in a set of panel estimates, each representing a weighted combination of different expert positions. SSHAC thus sees the TFI as performing an integration assisted by a group of experts who provide integration advice.

8. Thus, the TFI as facilitator structures interaction among the experts to create conditions under which the TFI's job as integrator will be simplified (e.g., either a consensus representation is formed or it is appropriate to weight equally the experts' evaluations of the knowledge of the technical community at large). In the rare case in which such simple integration is not appropriate, additional guidance is provided. In the main report, guidance is presented on two possible approaches involving (i) explicit quantitative but unequal weights (when it becomes obvious that using equal weighting misrepresents the community-as-a-whole); and (ii) "weighing" rather than "weighting", in cases when the experts themselves, acting as evaluators and integrators, find fixed numerical weights to be artificial, and when it is appropriate to represent the community's overall distribution in a less rigid way.

9. The SSHAC guidance gives special emphasis to the importance of an independent peer review. We distinguish between a participatory peer review and a late-stage peer review, and we also distinguish between a peer review of the process aspects and of the technical aspects for the more complex issues. We strongly recommend a participatory peer review, especially for the process aspects for the more complex issues. This paper details the pitfalls of an inadequate peer review.

APPENDIX B: Overview of Probabilistic Seismic Hazard Analysis

The SSHAC guidelines define a structured framework for developing the input to hazard analyses using expert assessment. This Appendix provides an introduction to hazard analyses in order to provide the context in which a SSHAC process would be applied. In particular it describes the uncertainties that the process is intended to identify, quantify, and incorporate into the analysis of seismic hazard. This Appendix has been included to provide a condensed introduction to the subject of seismic hazard analysis, and it makes no attempt to be a comprehensive guide to carrying out such analyses. The Appendix is primarily intended for readers who are not engaged in seismic hazard analyses but who wish to attain sufficient insight into the seismic hazard analysis procedures to follow the guidelines in this NUREG for their execution without needing to refer to other documents. The treatment of several topics is necessarily brief and for more detailed information the reader is directed to additional sources such as Reiter (1990), Thenhaus and Campbell (2003), McGuire (2004) and Bozorgnia and Campbell (2004b), as well as the references cited both in the remainder of this Appendix and in the sources noted. As well as being brief in overall coverage of a complex topic, this Appendix provides a treatment of different aspects of seismic hazard analysis that is somewhat uneven, with greater emphasis on ground-motion prediction models than on models for the characterization of seismic sources. The reason for this is that while seismicity is the driving factor in hazard analysis (the seismic hazard curve scales directly with the earthquake activity rate of the dominant seismic sources), the uncertainty in the ground-motion prediction tends to be the dominant contributor to the overall uncertainty.

The Appendix begins by providing an overview of how hazard assessment fits into the context of risk assessment in Section B.1. Section B.2 then presents a brief introduction to basic seismological definitions to provide the background to the description of probabilistic seismic hazard analysis (PSHA) in Section B.3. The discussion of PSHA is structured around the treatment of different uncertainties, so the section begins with the distinction between aleatory variability and epistemic uncertainty, before explaining how the former is integrated into the hazard calculations and the latter is handled using logic-trees. The relevance of the discussion of epistemic uncertainty is very clear, given that the SSHAC process is essentially a structured framework for building logic-trees for input to PSHA.

Although most of this Appendix deals with the basic elements of conducting PSHA, Section B.4 discusses the more complex issue of the degree to which PSHA results can be tested or validated. This discussion is included partly because considerable work has appeared on this topic in recent years and it is important for practitioners and users to be aware of the extent to which probabilistic hazard estimates can be verified, which is closely related to the understanding of the nature (and size) of the uncertainties discussed earlier in the Appendix.

The focus in this Appendix is primarily on the analysis of seismic hazard in terms of vibratory ground motion in rock, with only passing mention of the incorporation of site effects. However, because the SSHAC framework can equally be applied to the assessment of any geological hazard, the Appendix finishes with a brief overview of how the probabilistic hazard analysis framework can be adapted to other effects of earthquakes such as surface fault rupture and tsunamis.

B.1   Input to Risk-Informed Design and Evaluation

Deterministic approaches to hazard assessment, in which single values of all the required parameters are selected (based on a chosen earthquake scenario) and then a single value of the hazard levels to be used in design or review are calculated, are still in use around the world. In some countries, such approaches are even embedded in regulatory guidance. These deterministic approaches do have the advantage of being much less complex to apply than their probabilistic counterparts, and they are also easier to understand and interpret. However, in these guidelines the focus is almost exclusively on probabilistic approaches to hazard analysis, because these allow a more comprehensive treatment of uncertainties and are the basis for the NRC's current approach and guidance. Design against the maximum physically realizable cases of most types of loading[17] would actually be unfeasible, even for safety-critical structures. Instead, pragmatic engineering requires a rational approach to determination of the appropriate hazard levels to be used in design, and a way to select design loads that are below these highly unlikely extremes but still assure acceptable risk levels are not exceeded. Probabilistic seismic hazard analysis provides a rational procedure for determining an appropriate loading case.

Another important use of a hazard analysis is to define input parameters to risk calculations, whether it be for the evaluation of the risk in existing structures and facilities or for the risk-informed design of new structures. Risk is related to the potential impacts associated with failure of an individual structure, system, or component—or with the failure of a series of structures, systems, and components (SSCs)—as a result of an event, such as an earthquake. Risk can be quantified in terms of impact to human life and well-being, economic costs, interruption of services or production, or impact on the natural environment. Hazard analyses generally quantify loads imposed by natural events or activities external to the facility over which little control can be exerted. In these cases, risk can only be reduced by designing SSCs to have higher levels of seismic resistance (or relocating the project to a site of lower hazard). Such increases in seismic resistance have cost implications, so a balance is sought between the level of resistance provided (often referred to as *capacity*) and the level of loading that is expected to be exerted on the structure or facility (often referred to as *demand*). The basis for finding this balance will be the tolerable level of the consequences of failure (with respect to a specified performance target), which will clearly depend on the nature and the function of the facility.

For a nuclear power plant, the ultimate goal (but not the only one) is to prevent release of radioactive material resulting from damage to the reactor core. Risk is quantified both in terms of an adverse consequence (in this case damage to the reactor core) and the chance of this consequence occurring, expressed usually as an annual frequency or probability. Risks must be defined in terms of both consequence and rate of occurrence because this is ultimately the only transparent and technically robust basis for informed decisions regarding investing limited resources in the mitigation of different risks to society, to individuals, and to the environment.

If estimates of risk are to be defined in terms of consequences and their likelihood of occurring, the demand must be determined as a function of likelihood or rate, which can only be achieved through probabilistic analysis. The overall frequency of loss can be calculated from the convolution of a hazard curve (displaying the frequency of exceedance of different levels of demand) and a fragility curve (showing the probability of failing a specific performance criterion given particular levels of demand) for a given structure, system, or component. In order to fully analyze the risk to a nuclear power plant, the fragility curves for all of the SSCs must be determined and their interactions and dependencies must be defined in a systems analysis. Given this information, a probabilistic risk analysis (PRA) can develop an overall risk estimate. The process is illustrated schematically in Figure B-1. If greater capacity is provided through enhanced seismic design, then the fragility curve is modified (shifted to the right and/or made less steep) and consequently the risk (*i.e.*, the annual frequency of core damage) will be reduced.

---

[17] The maximum physically realizable cases of most types of loading associated with natural hazards are those that are possible, but are so rare as to have a nearly negligible probability of actually occurring.

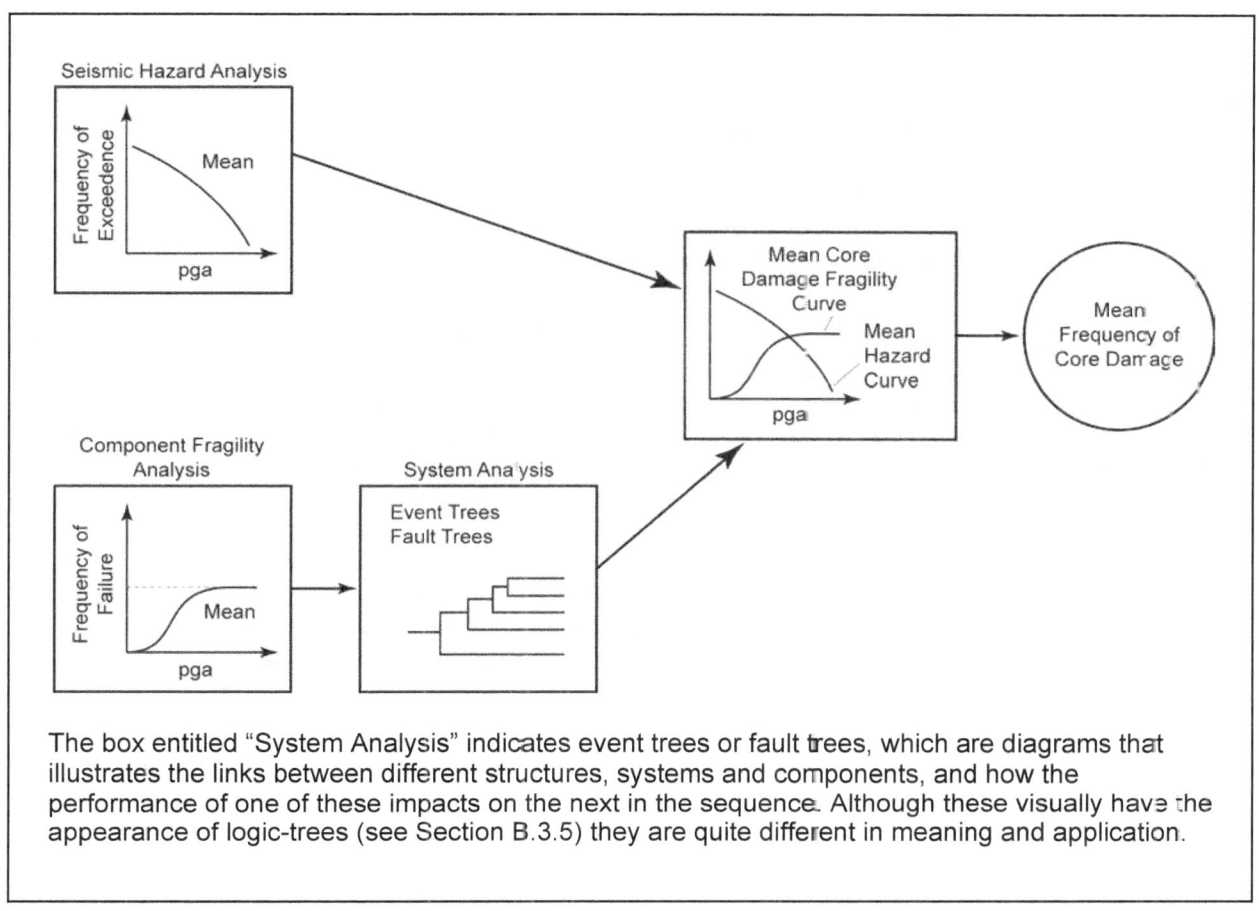

The box entitled "System Analysis" indicates event trees or fault trees, which are diagrams that illustrates the links between different structures, systems and components, and how the performance of one of these impacts on the next in the sequence. Although these visually have the appearance of logic-trees (see Section B.3.5) they are quite different in meaning and application.

Figure B-1. Risk-Assessment Methodology for Seismic Input (Kennedy, 1999).

This same technique can also be used in a risk-informed framework to determine the appropriate definition of design loads. This is done by working backwards from target failure probabilities and by including a requirement that adequate margins (required to meet risk-informed goals) are part of the design. Through such an approach, the appropriate basic input to nuclear reactor design can be assessed in terms of a loading level associated with a particular probability of exceedance. For the case of seismic loading, this is exemplified in the approach used to define the Safe Shutdown Earthquake (SSE) in ASCE 43-05 (ASCE, 2004), which is also embedded in Regulatory Guide 1.208 (USNRC, 2007). The approach is based on a simplification of the seismic risk calculation and the specification that design loads have annual frequencies of exceedance between 10^{-4} and 10^{-5}, depending on the slope of the hazard curve (Kennedy, 2007).

The requirement to define acceptable risk in terms of both performance-related criteria (defined by various unacceptable consequences to the public and the environment) and the expected rate of the occurrence of the unacceptable consequence is sufficient to require probabilistic assessment of hazard for nuclear facilities. Another compelling reason to adopt a probabilistic framework for hazard analysis is that it offers a rational and transparent framework for handling uncertainties, as discussed further in Section B.3.

The framework of probabilistic analysis can be adapted to any hazard, whether natural or anthropogenic. The specific case of earthquake-induced ground shaking is the main focus of this Appendix – as it was in the original SSHAC guidelines (NUREG/CR-6372) – but that does not limit the usefulness of probabilistic analysis and the SSHAC approach to this particular hazard. Section B.5 defines a general framework for probabilistic hazard analysis and then briefly explains its application to other geo-hazards.

B.2   Earthquake Processes, Hazards and Parameters

As noted above, the probabilistic analysis of seismic hazard in terms of ground shaking is the primary focus of this appendix and it is the hazard case used to illustrate the fundamentals of probabilistic approaches to hazard assessment. In order to introduce methods used for the assessment of this particular hazard more clearly, this section briefly introduces earthquake processes and the parameters that are used to quantify and characterize earthquakes and their effects.

With the exception of small volcanic earthquakes caused by the movement of magma and very deep subduction earthquakes that are associated with phase changes of the materials in the descending crust, earthquakes are caused by abrupt slip on geological faults. The sudden movement of a fault releases stresses in the surrounding crustal rocks, thereby releasing elastic strain energy that propagates away from the source in the form of elastic or seismic waves. These two processes, fault slip and the consequent radiation of seismic waves, give rise to all of the potentially destructive effects that earthquakes can inflict on the built environment. These hazards, and their relationship to the underlying earthquake processes, are illustrated in Figure B-2.

If the fault rupture extends to the ground surface, it will result in differential displacement of the surface, which poses an obvious and direct threat to any structure straddling the fault trace. The probabilistic assessment of surface rupture hazard is discussed in Section B.5.

If the fault is located offshore and the rupture is shallow enough to cause vertical displacement of the seabed, the sudden displacement of the seafloor and overlying water column will cause a pressure disturbance within the water column that will propagate away from the area of the earthquake source in the form a gravity wave called a tsunami. In the open ocean, tsunami waves travel at great speed[18] and have large wavelengths and small amplitudes. However, as such waves approach the shore and the water depth decreases, so too does the velocity of the wave and consequently the height of the wave increases to maintain the energy carried by the wave. The friction of the sea bottom in the near-shore environment also slows the bottom of the water column and creates a shoaling affect. As a result, tsunamis can cause great destruction in low-lying coastal areas through direct impact of the waves (which can reach heights of tens of meters), through inundation, and as a result of their ability to carry large quantities of debris that can batter the onshore built environment. Tsunamis can also be triggered by submarine slides and slumping, as indicated in Figure B-2, as well as volcanic phenomena and extraterrestrial impacts. The probabilistic assessment of tsunami hazard is discussed in Section B.5.

---

[18] The travel velocity of a tsunami wave is the square root of the product of the ocean depth and the acceleration due to gravity.

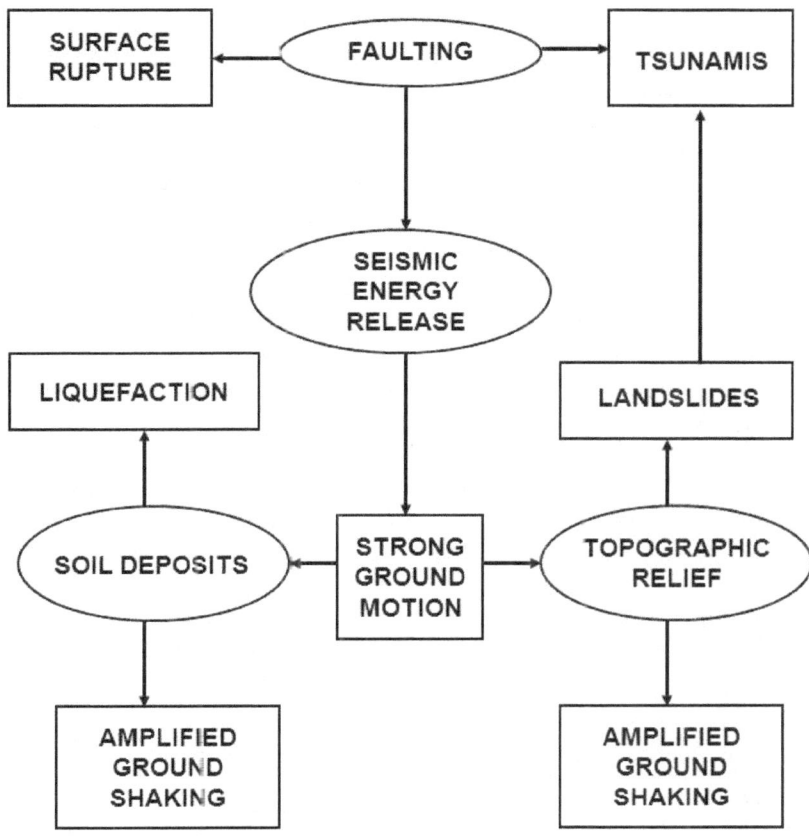

Figure B-2. Earthquake Hazards (In Squares) and Their Relationship to Earthquake Processes and Their Interaction with the Natural Environment (after Bommer and Boore, 2004).

The other hazards associated with earthquakes are all associated with the ground shaking caused by the passage of seismic waves radiated from the earthquake source. As indicated in Figure B-2, the magnitude of shaking at a particular site is influenced by the nature of the surface geology at the site, with softer soils generally leading to amplification of the incoming motion. Topographical features such as hills and ridges can also amplify the ground shaking In general, the majority of earthquake-induced damage within a region is due to the direct effects of strong ground shaking. However, at specific sites important secondary effects such as liquefaction or landslides can account for the majority of damage. Liquefaction is a phenomenon that occurs in saturated, cohesionless soils when ground shaking causes pore pressure to rise and reduce the effective stress to zero, thereby allowing the soil to temporarily behave as a liquid and undergo extensive deformations. The assessment of liquefaction hazard is briefly discussed in Section B.5. Secondary hazards can also have an important impact on many lifelines (communication systems, pipelines, etc.) (Bird and Bommer, 2004).

Another collateral hazard that can result from the effects of ground shaking on the natural environment is earthquake-induced landslides (Keefer, 1984; Rodriguez et al., 1999). The assessment of this hazard is also discussed briefly in Section B.5.

Focusing on the primary hazards of surface fault rupture and ground shaking, a few key parameters will now be defined that will be used in later sections when discussing the assessment of hazards. The geometry of a geological fault rupture is defined by its orientation and dimensions, for which the simplifying assumption is usually made that the fault can be represented by a rectangular plane. The direction of the line that the fault traces on the surface (or in the case of sub-surface rupture planes, the trace that *would* occur on the surface if the rupture propagated to the surface) is known as the strike. The strike is defined by the compass azimuth of the trace and is oriented such that the fault dips to the right. The dip of the fault is the angle of the fault plane measured from horizontal. Fault ruptures are classified according to the sense of slip between the two sides of the fault. The first distinction is between those that move horizontally (strike-slip) and those that move vertically (dip-slip). Oblique ruptures are those that involve both horizontal and vertical displacement. Strike-slip faults are classified as left-lateral or right-lateral depending on the direction of movement of the opposite side of the fault relative to the observer. Dip-slip faults are classified as normal if the hanging wall (the crustal block above the dipping fault plane) moves downwards with respect to the foot wall, and reverse if the hanging wall moves upwards. Thrust faults are a special case of reverse faults with very low dip angles.

The size of an earthquake is defined by the amount of seismic energy released by the fault rupture. This energy can be measured by the seismic moment, M_o, which is defined as the product of the fault rupture area, A, the average slip on the fault plane, u, and the rigidity of the crust, μ (usually taken as 3.3×10^{10} N·m^{-2}):

$$M_o = A \cdot u \cdot \mu$$

Equation B.1

The moment magnitude (Hanks and Kanamori, 1979) is defined by the following equation when the seismic moment is given in N·m:

$$M_w = \frac{2}{3} log_{10}(M_o) - 6.0$$

Equation B.2

In practice, seismic moment for a specific earthquake is determined from the Fourier amplitude spectra of seismograms rather than from Equation (B.1). Earthquake magnitude can also be determined from the amplitude of instrumental recordings using a variety of other scales, including local magnitude, M_L, and teleseismic scales such as body-wave magnitude, m_b, and surface-wave magnitude, M_s. Each magnitude scale is based on a different part of the seismic spectrum because each is based on recordings from seismographs with different dynamic characteristics. Therefore, the magnitude of an earthquake will not generally have the same value on all scales as a result of the different natural frequencies of the seismographs on which the recordings are made. Many empirical relationships have been derived between different magnitude scales, an overview of which is presented by Utsu (2002).

The dimensions of the fault rupture, and the slip on the fault rupture, increase exponentially with earthquake magnitude (e.g., Wells and Coppersmith, 1994; Hanks and Bakun, 2002; Leonard, 2010; Blaser *et al.*, 2010; Strasser *et al.*, 2010). Median predicted rupture lengths and fault slips from the empirical equations of Wells and Coppersmith (1994) for strike-slip earthquakes are shown in Figure B-3.

The point on the fault at which rupture is initiated is known as the focus or hypocenter. The location of the hypocenter is described by the geographical coordinates of the epicenter and the focal depth. The epicenter is the vertical projection of the hypocenter onto the Earth's surface.

The focal depth is the vertical depth in kilometers from the epicenter at the Earth's surface to the hypocenter that marked the point of initiation of the earthquake.

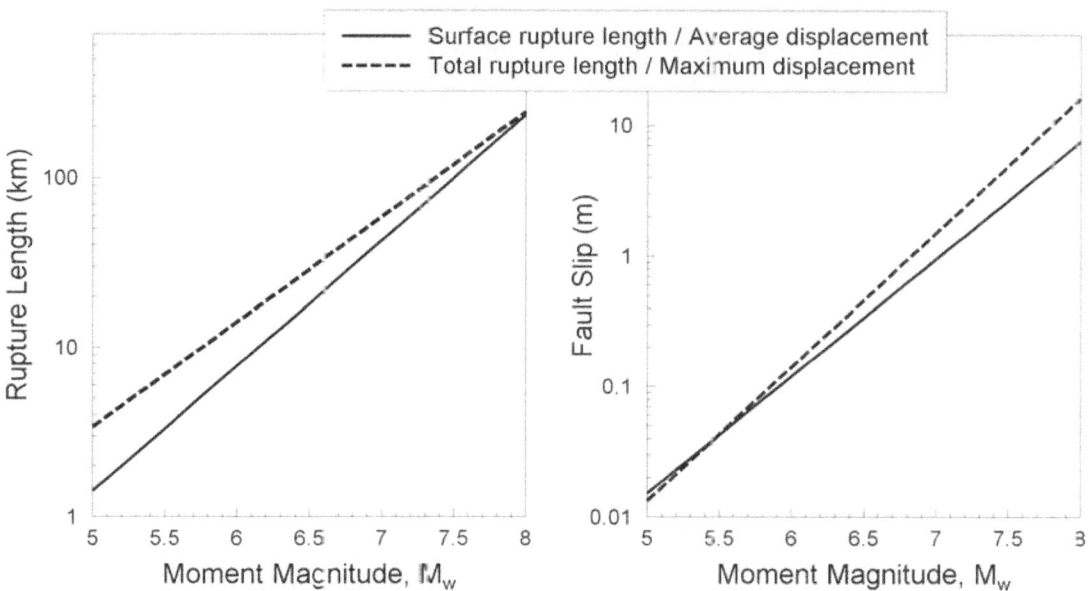

Figure B-3. Median Rupture Lengths (Left) and Fault Slip (Right) as a Function of Moment Magnitude, from the Strike-Slip Model of Wells and Coppersmith (1994) for Crustal Earthquakes.

The strength of shaking at any location during an earthquake will depend primarily on the size of the earthquake, how far the site is located from the earthquake source, and the type of geologic materials that exist at the site. The source-to-site distance can be measured with respect to the epicenter, r_{epi}, or the hypocenter, r_{hyp}. However, it is clear from Figure B-4 that for larger earthquakes the rupture dimensions are such that this approximation of a point source becomes inadequate, especially for sites close to the source that will experience strong shaking. For such circumstances, distances can be measured directly from the closest point on the fault rupture, r_{rup}, the closest point on the surface projection of the fault rupture, r_{JB} (also referred to as the Joyner-Boore distance), or r_{seis}, which is similar to r_{rup} but measures the distance to the closest point on the fault rupture within the seismogenic layer of the crust. Several of these distance metrics are illustrated in Figure B-4. The choice of which distance metric to use is not straightforward because none of the measures that account for the extension of the fault rupture have been proven to consistently perform better in terms of predictions. Each of the measures is an attempt to capture the distance over which the amplitude of motion decays (through geometrical spreading and anelastic attenuation) with increasing separation from the source of energy release, which in reality is a volume of the Earth's crust around the fault rupture. Although these distance measures can reflect the fact that the source of energy release is extended rather than a single point, none of them account for uneven distribution of energy release along the length of the fault rupture.

The nature and amplitude of the ground shaking at any particular site can be modified, as noted previously, by near-surface layers at the site, particularly if these are softer than the underlying geologic materials. For the purposes of predicting the influence of the near-surface geo-materials on the surface motions, it is common to characterize the site by the average shear-wave velocity over the uppermost 30 meters, $V_{s,30}$, which can be assigned based on general site classification[19] or based on site-specific geotechnical information. Nuclear facilities, however, require highly detailed characterization of the underlying soil column and analysis of how the stratigraphy transmits seismic energy, as well as its impact on overlying structures.

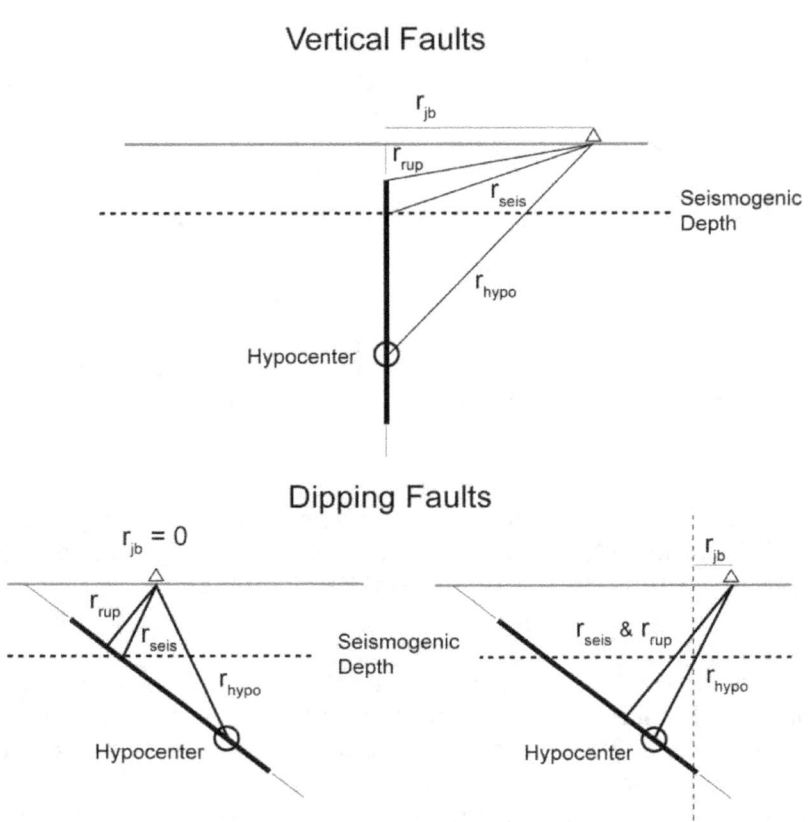

Figure B-4. Distance Metrics for the Location of a Site with Respect to the Source of an Earthquake (Abrahamson and Shedlock, 1997).

The strong shaking experienced at sites resulting from earthquakes is recorded on instruments called accelerographs. The recordings of ground acceleration against time, known as accelerograms (Figure B-5), are usually obtained in two orthogonal directions horizontally and the vertical direction. After suitable processing to compensate for noise in the recordings (e.g., Boore and Bommer, 2005), the acceleration can be integrated over time to obtain the velocity time-history (also shown on Figure B-5). The simplest parameters that can be defined to characterize the nature of the shaking are the peak absolute values from these time series, namely the peak ground acceleration (PGA) and the peak ground velocity (PGV).

---

[19] For non-critical facilities only

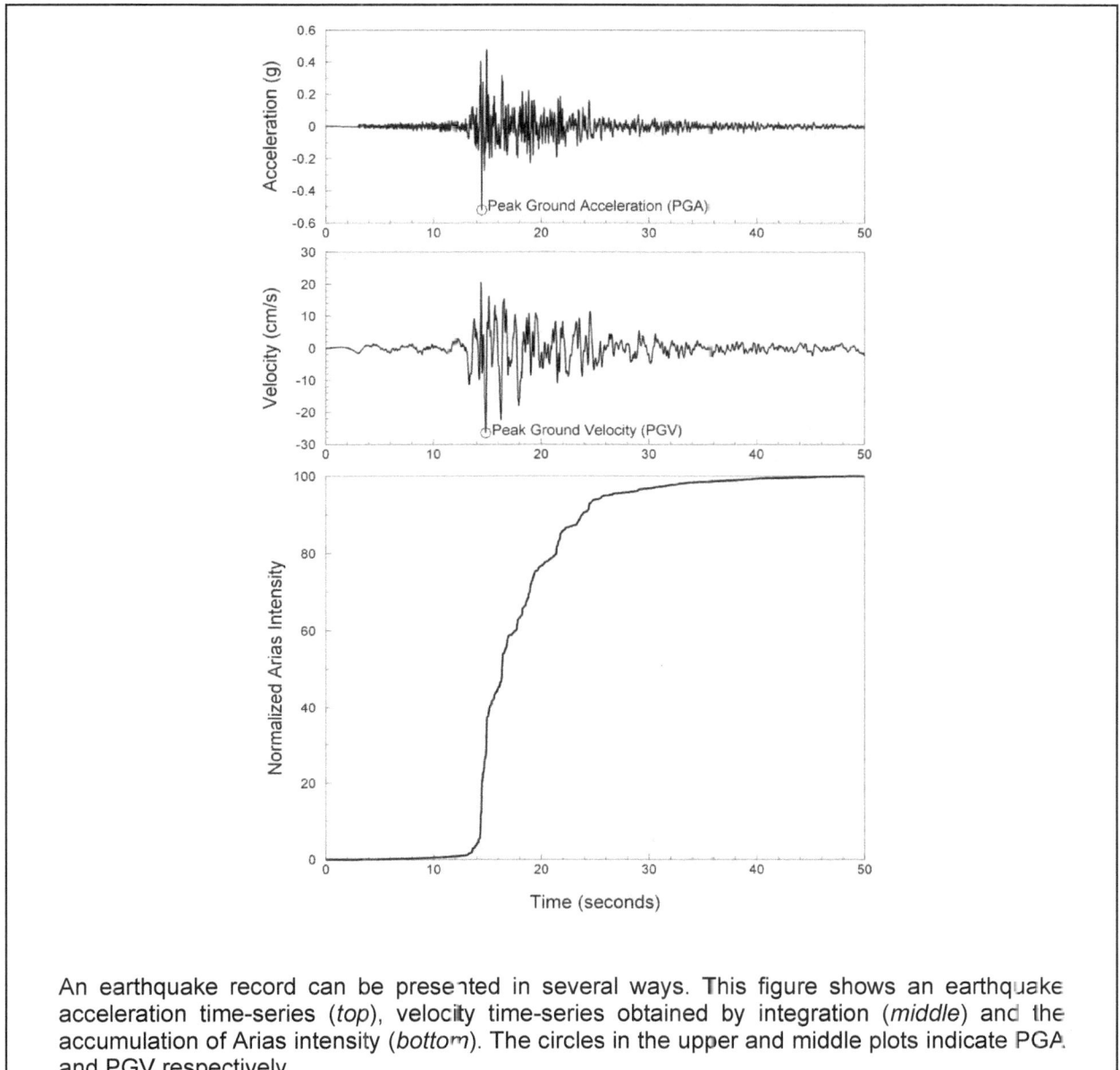

An earthquake record can be presented in several ways. This figure shows an earthquake acceleration time-series (*top*), velocity time-series obtained by integration (*middle*) and the accumulation of Arias intensity (*bottom*). The circles in the upper and middle plots indicate PGA and PGV respectively.

Figure B-5. An Earthquake Acceleration Time Series, and Velocity Time Series, and the Accumulation of Arias Intensity.

Another parameter that can be calculated from the accelerogram is the Arias intensity (Arias, 1970), which is defined by the following equation, in which $a(t)$ is the acceleration time history, t_{max} the duration of the complete recording, and g the acceleration due to gravity:

$$AI = \frac{\pi}{2g} \int_{0}^{t_{max}} a^2(t)\,dt$$
Equation B.3

The Arias intensity is a measure of the energy in the motion; the final frame in Figure B-5 shows the build-up of Arias intensity over time, expressed as a proportion of the final value.

The duration of the shaking can be measured in many different ways (Bommer and Martinez-Pereira, 1999), usually either on the basis of the times at which an acceleration threshold is exceeded or the interval over which a specified proportion of the total Arias intensity is accumulated. Alternatively, the number of cycles of motion can be measured. However, for this parameter there are also a large number of available definitions (Hancock and Bommer, 2005). Surprisingly, duration and numbers of cycles are very poorly correlated, regardless of the definitions employed (Bommer et al., 2006).

Another parameter of relevance to nuclear applications is the cumulative absolute velocity (CAV). Originally proposed by Reed and Kennedy (1998), it is most directly defined as the integral of the acceleration over the length of the record. However, its use in nuclear applications has subsequently been modified such that it is used as a "CAV filter" (see Section 5.9). The CAV filter has been defined so that the integral only includes those 1-second time windows in the record in which the maximum acceleration is greater than 0.025g:

$$CAV = \sum_{i=1}^{N} H(PGA_i - 0.025) \int_{t=t_i}^{t_{i+1}} |a(t)| dt \qquad \text{Equation B.4}$$

where PGA_i is the maximum acceleration in the ith 1-second time window, and $H(x)$ is the Heaviside function (equal to unity for $x > 0$, zero otherwise).

The single most important characterization of earthquake strong-motion from an engineering perspective is the response spectrum, which shows the maximum response of a single-degree-of-freedom (SDOF) damped oscillator to a particular excitation. The natural frequency, f, of vibration of an SDOF oscillator of mass m and flexural stiffness k is given by:

$$f = \frac{1}{2\pi} \sqrt{\frac{k}{m}} \qquad \text{Equation B.5}$$

The oscillator can alternatively be characterized by its natural period of vibration, which is simply the reciprocal of the natural frequency. Figure B-6 shows the acceleration response spectra for SDOF oscillators with various levels of critical damping (which is the damping that will cause a displaced oscillator to return to its original position without vibration) under excitation by the accelerogram in Figure B-5. Note that at high frequencies, regardless of the damping level, the acceleration response becomes constant, and equal to the PGA of the accelerogram. A very high frequency of vibration implies that the oscillator is extremely stiff and hence it does not move relative to the base (i.e., it does not vibrate but rather moves in unison with the ground). In this case, the maximum acceleration experienced by the mass is equal to the maximum acceleration of the base. Mathematically, for f to be high, k must be very large (Equation B.5). The period of vibration is the reciprocal of the frequency, and this becomes zero when the stiffness is very high, in which case the spectral acceleration is equal to PGA.

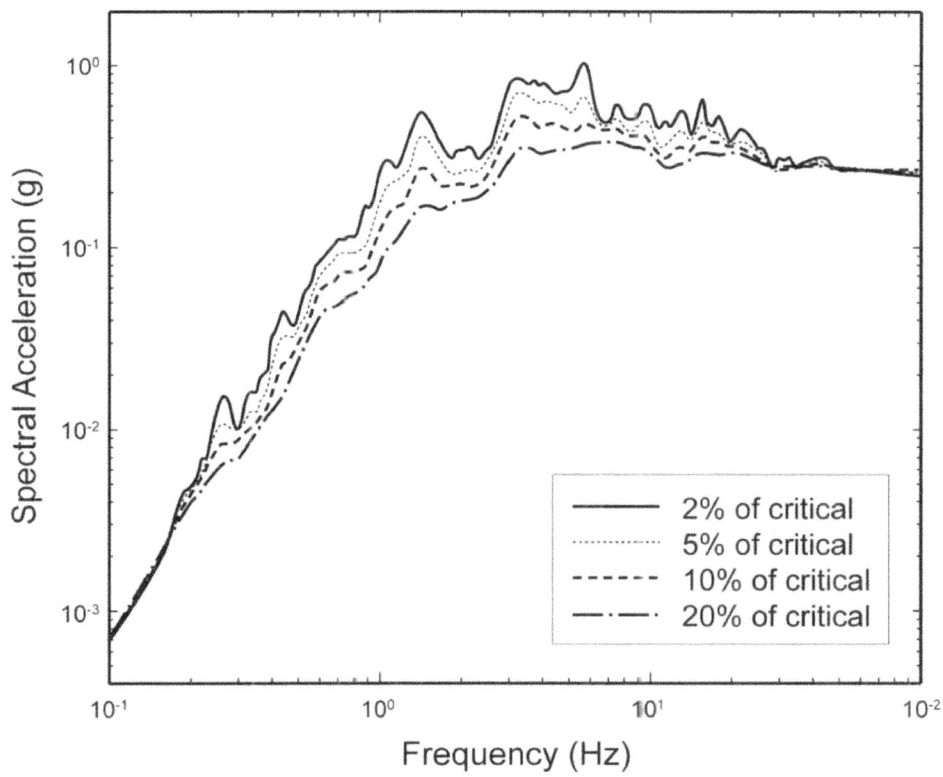

Figure B-6. Acceleration Response Spectra (for 4 Values of Critical Damping) of the Accelerogram in Figure B-5.

Ground shaking can also be characterized by intensity, which is a site-specific qualitative index based on field observations of earthquake effects and reports of how the shaking was experienced by people. Intensity is reported as an integer value (usually in Roman numerals) assigned on the basis of the modal[20] observation of the strength of shaking in a given location.

Lower degrees of intensity are assigned primarily on how widely and how strongly the shaking is felt by people, whereas higher degrees of intensity are assigned on the basis of the degree of damage experienced by buildings of different vulnerabilities. A number of intensity scales exist, including the 12-degree Modified Mercalli scale used in the Americas and the very similar European Macroseismic scale used in Europe and surrounding areas. A few countries, including Italy and Japan, have their own scales with fewer degrees.

Intensity has many uses, particularly for quantifying and characterizing the effect of earthquakes that occurred prior to the introduction of seismographs (c. 1900). The use of intensity information enables locations and magnitudes to be estimated for pre-instrumental earthquakes. However, intensity cannot be used directly as input to engineering analysis and design. Empirical relationships have been derived between degrees of intensity and instrumentally-

---

[20] In this case the term "modal" refers to the statistical mode (i.e., most common) of the intensity observations.

recorded ground-motion parameters (Allen and Wald, 2009), but there is always a very large degree of scatter in such relationships (Figure B-7).

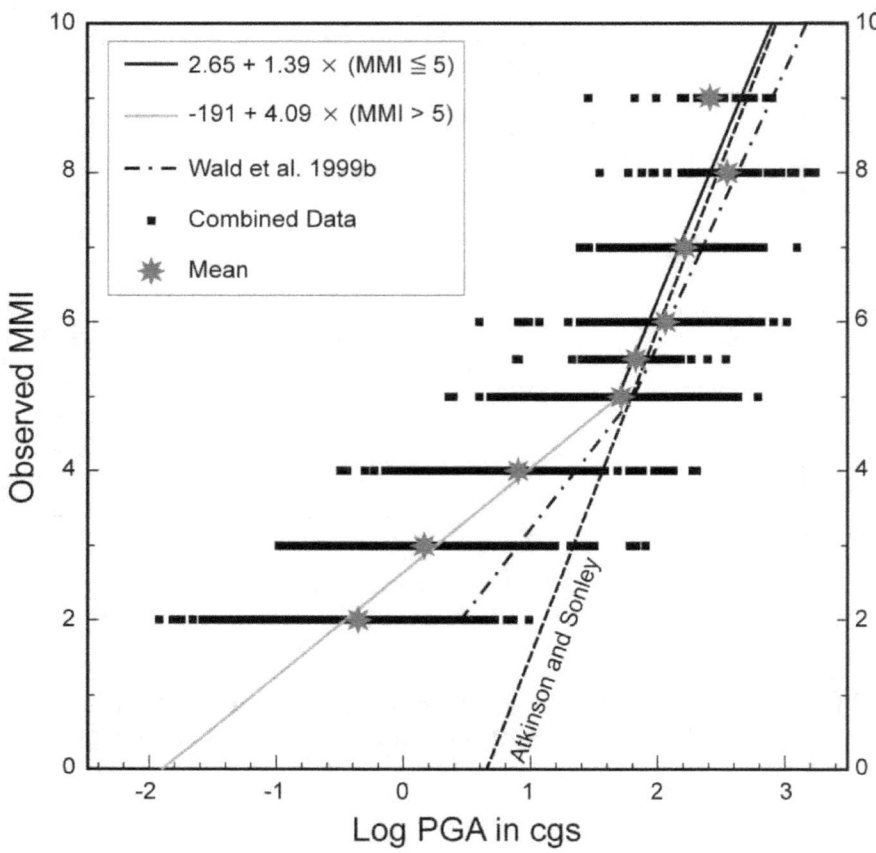

Figure B-7. Relationships Between PGA and Observed Modified Mercalli Intensity (Atkinson and Kaka, 2007).

B.3   PSHA and the Nature of Uncertainties

This section provides an overview of the basic elements of probabilistic seismic hazard analysis (PSHA), in terms of ground motions in rock (i.e., no site response or soil structure interaction is incorporated into the discussion). The information presented herein is only intended as an introduction in order to provide the reader with the background to understand how the SSHAC procedures fit into a site-specific assessment of seismic hazard. Greater detail on the specific details of conducting a PSHA can be found in the original SSHAC guidelines (NUREG/CR-6372) and other references noted at the start of this Appendix. Guidelines on conducting a site-specific PSHA to determine ground shaking levels to be used in the design of nuclear power plants are provided in Regulatory Guide 1.208 (USNRC, 2007).

The above references provide useful insight into the basic elements of PSHA and the overall process to calculate seismic hazard curves for different ground-motion parameters and the uniform hazard spectrum. However, they do not provide a complete formulation, including all the specific details regarding compatibility, adjustments and refinements, which need to be adapted to the elements of any individual study as a result of the variability of the quantity and quality of available information for a particular site. Additionally, very detailed step-by-step guidance for the execution of a site-specific PSHA has not been produced in a single volume because the state-of-the-art in this field is evolving very rapidly and any attempt to provide such a manual would quickly become outdated. In discussing the overview of PSHA that is provided in Chapters 4 to 6 of volume 1, the SSHAC Guidelines state that the provided *"formulation should not be viewed as an attempt to 'standardize' PSHA in the sense of freezing the science and technology that underlies a competent PSHA"* (NUREG/CR-6372). Because the state-of-the-art in PSHA is advancing so rapidly, and the advances are often reported in the technical literature with some delay (and even then in summary form), it is important to recognize that staying informed of the latest developments in probabilistic seismic hazard analysis requires dedication of considerable time and effort. Practical experience and familiarity with the relevant research communities are therefore vital attributes for those charged with conducting PSHA studies for critical facilities.

B.3.1 The Nature of Uncertainties

The primary objective of a PSHA is to produce a seismic hazard curve for each ground-motion parameter of interest. These curves express the relationship between different values of the parameter of interest and the associated annual frequency of that value being exceeded. However, an equally important aspect is that the uncertainty associated with each curve calculated is assessed. As stated in the original SSHAC Guidelines:

> *"The most important and fundamental fact that must be understood about a PSHA is that the objective of estimating annual frequencies of earthquake-caused ground motions can be attained only with significant uncertainty"* (NUREG/CR-6372).

Two fundamental types of uncertainty can be identified; and a variety of terms have been used for these two types of uncertainties in different fields. Some have argued that in classical decision theory, the distinction between the two types of uncertainty is unimportant and that only the total uncertainty matters (e.g., Veneziano et al., 2009). Hofer (1996) presents an interesting discussion of cases where the separation of uncertainties is or is not useful. For practical application of PSHA, it is assumed in these guidelines that the distinction is important because the way in which the two classes of uncertainty are handled in seismic hazard analysis is very different.

The two classes of uncertainty have traditionally been referred to by the simple terms *randomness* and *uncertainty*. However, over the years, the usage of these terms has become lax and they have often been employed interchangeably, thereby blurring the distinction. For this reason, in PSHA, new terms have been introduced to encourage greater clarity when treating the uncertainties associated with the assessment of seismic hazard. The two types of uncertainty are now referred to as *aleatory variability* and *epistemic uncertainty* (NUREG/CR-6372); the original SSHAC guidelines referred to these as aleatory uncertainty and epistemic uncertainty, but the terminology adopted herein makes the distinction even clearer.

Aleatory variability is the inherent randomness in a process. Its name derives from *alea*, the Latin word for dice. This type of uncertainty has also been referred to as randomness, variability, irreducible uncertainty, inherent uncertainty, natural variability, type-A uncertainty, and stochastic uncertainty. In theory, aleatory variability cannot be reduced, because it is interpreted as representing an inherent quality of Nature. In terms of the practice of PSHA, it is more useful to consider aleatory variability explicitly in association with a particular model of Nature defined in the study because it is reflected in the calculation of random variability of observations specifically with respect to the model predictions. For this reason, it may be helpful to think of this type of uncertainty as "apparent" random variability, in which it is possible that some epistemic uncertainty has been included. Then, as more sophisticated models are developed that incorporate a greater number of explanatory variables—and perhaps a more complex functional form—the aleatory variability can be transformed to epistemic uncertainty on the explanatory variables.

Aleatory variability is represented by a probability distribution for known parameters. Once we have a model for a particular earthquake process, it is possible to define the range of values that the predicted variable is expected to take. However, the aleatory variability means that we do not know which value a parameter will take in any particular instance. An example of this is that we may define the expected distribution of focal depths of earthquakes in a given region based on well-determined depths in the earthquake catalog; but we have no way of knowing at what depth within the seismogenic layer any individual future earthquake may occur. Similarly, we may be able to determine the average recurrence rate for earthquakes of different magnitudes in an earthquake catalog, but this does not mean that we can know the magnitudes of the earthquakes that will occur in the next decade or the size of the next earthquake.

Epistemic uncertainty arises because our analyses are conducted using models (mathematical and conceptual constructs of reality) rather than directly on real systems. The adjective epistemic is derived from the Greek ἐπιστήμη (epistêmê) meaning knowledge; and this uncertainty reflects our lack of knowledge regarding earthquake processes, both in general and in a specific location. Epistemic uncertainty has also been referred to as reducible uncertainty, subjective uncertainty, model form uncertainty, ignorance, type-B uncertainty, and specification error. Unlike aleatory variability, epistemic uncertainty can—in theory at least—be reduced through the acquisition of additional data.

Whereas aleatory variability is always represented by probability distributions, there are more options for the representation of epistemic uncertainty. In PSHA, it is increasingly common for epistemic uncertainty to be expressed through the use of a logic tree. The SSHAC process itself is primarily concerned with the capture of epistemic uncertainty, although both types of uncertainty must be fully expressed in the PSHA. Because epistemic uncertainty is the central focus of these guidelines, it is discussed in more detail in Section B.3.4 below.

In closing this introductory discussion, it can be noted that Elms (2004) identified a third category of uncertainty, to which he gave the name *ontological uncertainty*. It refers essentially to the unforeseen, in other words to events outside the range of possibilities considered physically possible when modeling the epistemic uncertainty. The definition of ontological uncertainty is somewhat academic and of little practical applicability. However, its existence is a reminder that there is always a danger, even a tendency in human behavior, of underestimating epistemic uncertainty.

B.3.2 Models for Aleatory Processes

In PSHA, the objective is to calculate the annual rate at which different levels of a specified ground-motion parameter will be exceeded at a particular site. This depends on two fundamental elements: a spatial and temporal model for earthquake occurrence and a model for the prediction of ground-motions at the site as a result of any particular earthquake.

As discussed in Section B.2, the basic measure of earthquake size is magnitude (preferably moment magnitude, which is directly related to seismic moment, the best measure of the size of earthquakes). The model for earthquake occurrences in the region around the site of interest needs to quantify the expected distribution of events of different magnitudes. As stated by Cornell (1968) in the original formulation of PSHA,

> "In the determination of the distribution of maximum annual earthquake intensity at a site, one must consider not only the distribution of the size (magnitude) of an event, but also its uncertain distance from the site and the uncertain number of events in any time period."

In other words, PSHA requires a model for the spatial and temporal distribution of earthquakes of different magnitudes. The spatial model, which represents where future earthquakes could be expected to occur, is composed of a model of seismic source zones. These zones may be broad areas or may be linear sources where they correspond to a single active fault (Figure B-8). The most common assumption regarding the spatial distribution of future earthquakes is that they are equally likely anywhere within each source zone, although variations on this approach can easily be invoked. It is also possible to dispense with area source zones completely and use the density of epicenters of recorded earthquakes, spatially smoothed to reflect the possible variations in future locations, as distributed sources (Frankel, 1995; Woo, 1996).

A model for the distribution of earthquakes of different magnitude (*i.e.*, their average rates) within each seismic source is given by a recurrence model. Generally the simplest form of recurrence model used in PSHA is the Gutenberg-Richter relationship, which is expressed as:

$$log_{10} N(m) = a - bm$$

Equation B.6

where N is the number of earthquakes of magnitude m or greater per year, and a and b are constants (Figure B-9). A temporal model for earthquake occurrence can be obtained by combining the recurrence relationship with a probability distribution, such as the widely-used Poisson distribution or the Brownian Passage Time (BPT) model (e.g., Matthews *et al.*, 2002).

In practice, an upper bound must be imposed on this distribution. The upper bound chosen reflects the largest earthquake that is considered possible within the particular seismic source, m_{max}. The parameter m_{max} is often defined as a distribution, rather than a single value, to address the associated uncertainty. A lower limit of magnitude, m_{min}, is also generally considered but this reflects the smallest earthquake considered as being of engineering significance. The value of m_{min} for a particular study is imposed to prevent small earthquakes producing non-damaging motions from inflating the hazard estimate (EPRI, 1989).

Rather than abruptly truncating Equation (B.6) at the upper limit, an exponential form of the recurrence relationship is generally used:

$$N(m) = v_{m_{min}} \left(\frac{e^{-\beta(m-m_{min})} - e^{-\beta(m_{max}-m_{min})}}{1 - e^{-\beta(m_{max}-m_{min})}} \right)$$

Equation B.7

where $v_{m_{min}}$ is the annual rate of earthquakes of magnitude m_{min} and greater. The parameter β is equal to the *b*-value in Equation (B.6), multiplied by *ln*(10).

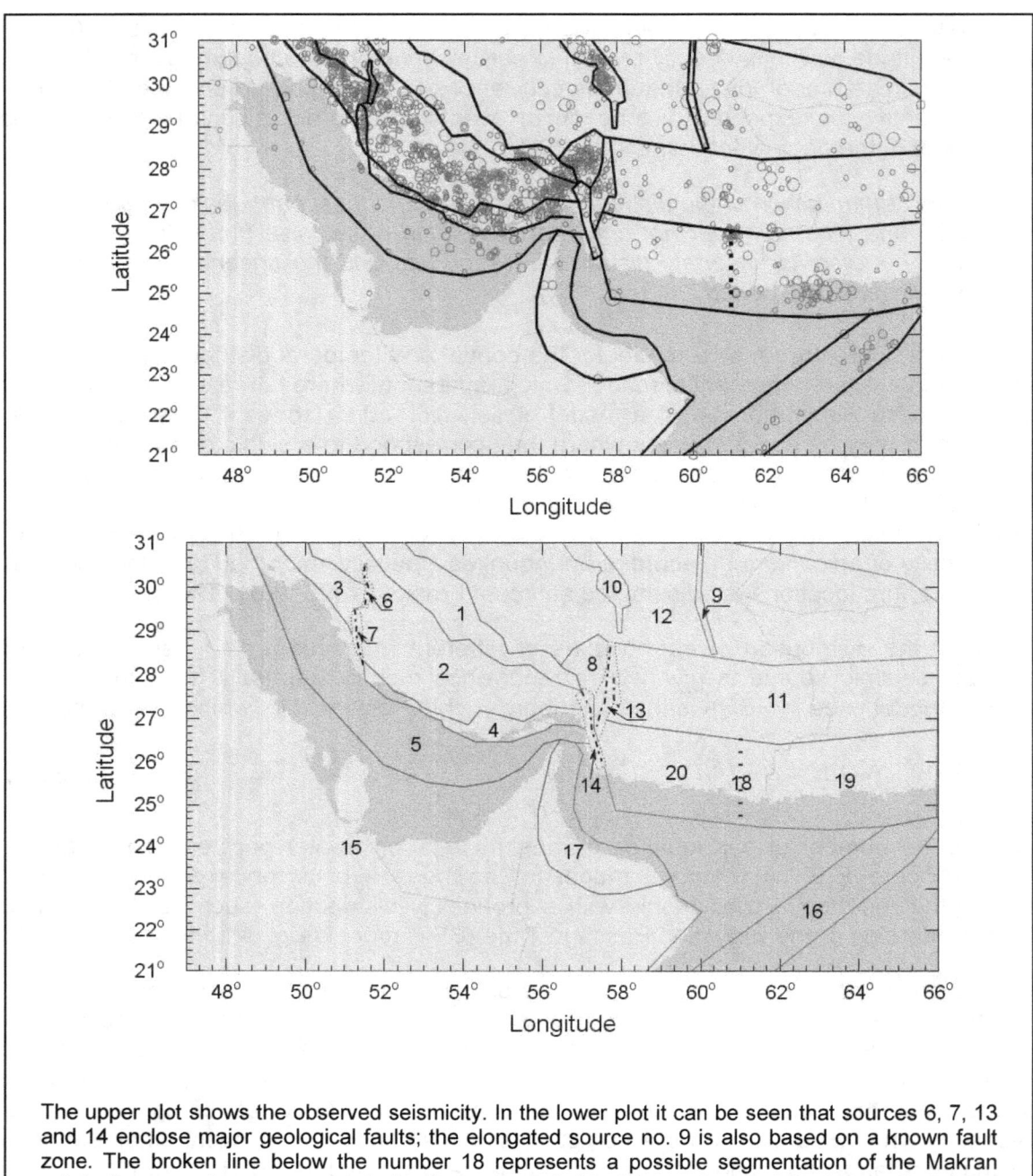

The upper plot shows the observed seismicity. In the lower plot it can be seen that sources 6, 7, 13 and 14 enclose major geological faults; the elongated source no. 9 is also based on a known fault zone. The broken line below the number 18 represents a possible segmentation of the Makran subduction zone, because some studies have concluded that the eastern and western portions have different seismogenic potential; one model (18) includes all of the Makran as a single seismic source, whereas another distinguishes two separate zones (19 and 20).

Figure B-8. Seismic Source Zones Defined for PSHA for the United Arab Emirates (Aldama-Bustos *et al.*, 2009).

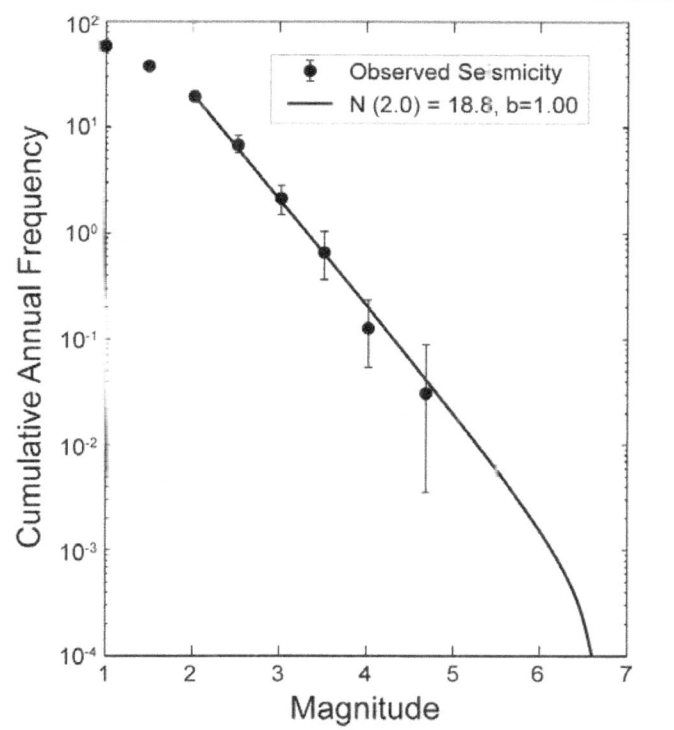

The dots denote mean annual frequency of observed earthquakes with magnitude greater than or equal to the central value of the magnitude interval and the vertical error bars denote the 90% confidence interval on the cumulative rate of observed earthquakes after correction for temporal completeness (NUREG/CR-6372).

Figure B-9. Example of the Observed Seismicity Within an Area Seismic Source and the Doubly-Truncated Exponential Recurrence Relationship Found Using the Maximum Likelihood Technique.

In addition to the doubly-truncated Gutenberg-Richter relationship defined in Equation B.7, an alternative model is sometimes used that combines Equation B.7 for smaller magnitude events and then a "characteristic earthquake" model for large-magnitude earthquakes. This relationship is used for individual geological faults when the fault is known to produce large earthquakes with similar characteristics on a quasi-periodic basis that has a higher frequency (as determined from geological data) than would be predicted by extrapolation of the recurrence equation from the small events (Schwartz and Coppersmith, 1984; Youngs and Coppersmith, 1985; Wesnousky, 1994). For fault sources, there is still debate as to whether the Gutenberg-Richter or characteristic earthquake recurrence model should be used (Parsons and Geist, 2009). Indeed, the better choice is likely to differ based on the actual behavior of each fault, which is not always well understood. In the case where the smaller earthquakes that conform to the Gutenberg-Richter relationship are below the threshold considered in PSHA, the characteristic model becomes the simpler maximum magnitude model. These three basic recurrence models are illustrated in Figure B-10.

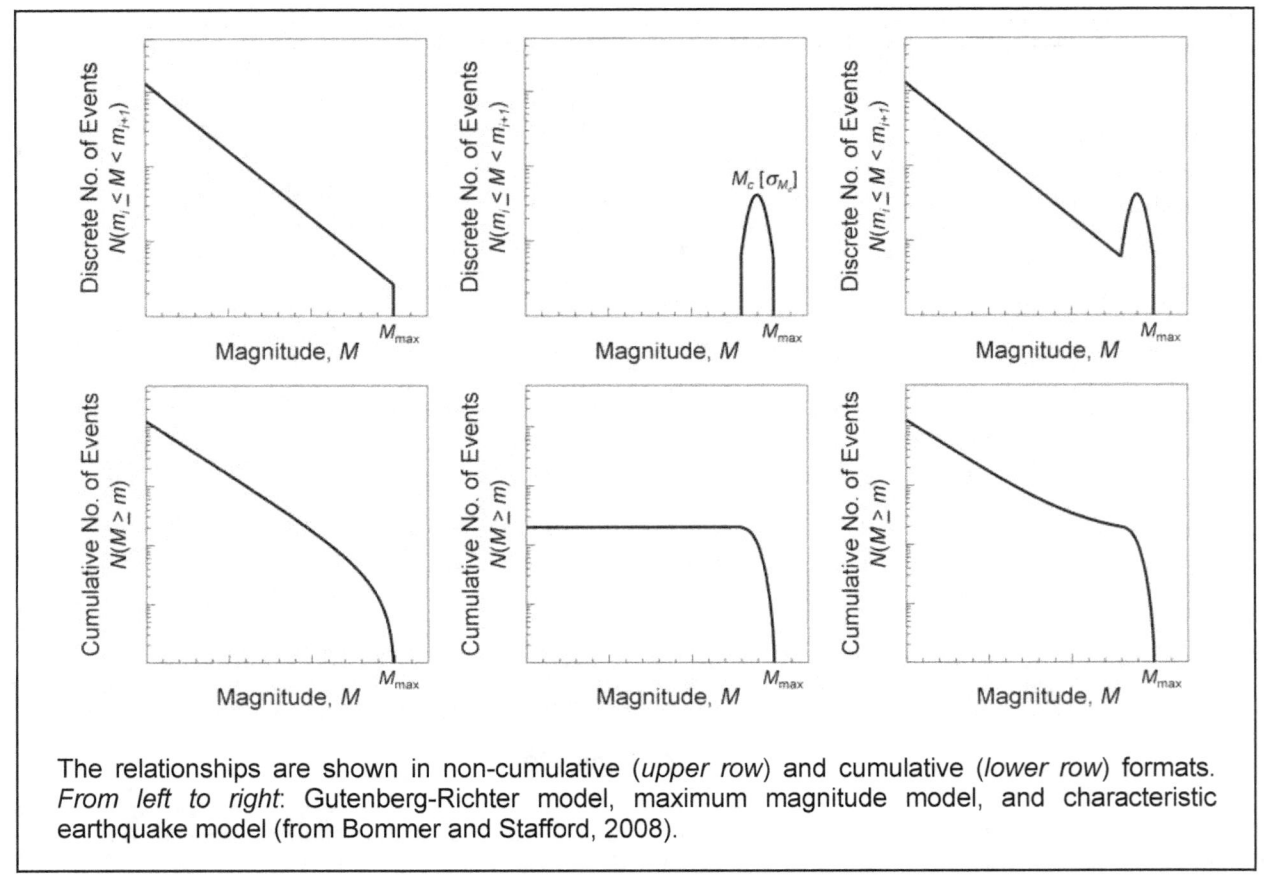

The relationships are shown in non-cumulative (*upper row*) and cumulative (*lower row*) formats. *From left to right*: Gutenberg-Richter model, maximum magnitude model, and characteristic earthquake model (from Bommer and Stafford, 2008).

Figure B-10. Typical Forms of Earthquake Recurrence Relationships.

In PSHA, the model for the spatial and temporal distribution of future earthquakes of different magnitudes is referred to as a seismic source characterization (SSC). For each earthquake scenario (i.e., scenario magnitude and location), a model is also required to predict the values of the chosen ground-motion parameter at the site in question. This is the counterpart to the SSC and is referred to as ground motion characterization (GMC). In practice this will usually include propagation of bedrock motions up to the surface, and foundation embedment and soil-structure analyses also come into play. However, as has already been noted, the focus here is only on rock motions, for simplicity. Empirical ground-motion prediction equations[21] (GMPEs) are derived through regression analysis on recorded values of the parameter of interest. GMPEs provide a distribution of the ground-motion parameter of interest based on explanatory variables that characterize the earthquake source, the travel path from the source to the site, and the nature of the site itself. The models generally include relatively few explanatory variables. Although the parameters used often include earthquake magnitude, style-of-faulting for the source (e.g., strike-slip, reverse, normal, or oblique), and distance between source and site of interest (using one of many definitions, as illustrated in Figure B-4), other explanatory variables can also be included. The influence of geologic materials that exist at the site of interest will

---

[21] Ground motion prediction equations were formerly known as "attenuation relationships", but this name fails to capture the fact that the equations model scaling of ground motions with magnitude as well as attenuation with distance.

usually be classified in terms of $V_{s,30}$ as noted previously in Section B.2, as well as depth to bedrock in some models.

Reviews of the state-of-the-art in empirical ground-motion prediction have been published at various times. These publications include Idriss (1978), Campbell (1985), Joyner and Boore (1988) and Douglas (2003, 2011), among others. The empirical GMPEs produced in the Next Generation of Attenuation (NGA) project (Power et al., 2008), which represent the state-of-the-art in empirical ground-motion modeling at the time of writing, include a number of additional explanatory variables to represent the influence of factors such as depth of embedment of the fault rupture, hanging wall effects, and depth of sediments (Abrahamson et al., 2008).

A very important point to stress is that the intention of a GMPE is not to provide the best model for the observations on which it is based but rather to provide stable predictions of ground motions for future earthquakes. The important distinction here is that the latter will include earthquake scenarios that are not represented in the dataset of recordings. Using a complex functional form that provides a marginally improved fit to recorded ground-motion amplitudes but does not extrapolate well to higher magnitudes or shorter distances than those represented in the database, is not desirable.

When the relatively simple functional forms used for GMPEs are fit to datasets of observations from strong-motion instruments, there is invariably a large degree of scatter in the data around predicted value (Figure B-11). This may reflect both inherent variability in the ground motion and the influence of parameters that are not included in the model, although the reductions in the scatter achieved by adding parameters beyond magnitude, distance and site classification tend to be very modest. The regressions are generally performed on the logarithm of the ground-motion parameter values, because it is found that the residuals (the difference between observed and predicted values of the ground-motion parameter) then conform to a standard normal distribution. The distribution of the residuals can therefore be fully characterized by a normal distribution with zero mean and a standard deviation, σ. A general equation for the prediction of a ground-motion parameter Y can then be expressed in the following form:

$$logY = f(X_1, X_2, ..., X_N) + \varepsilon \sigma$$

Equation B.8

where X_i are the N explanatory variables (magnitude, distance, etc.) in the equation and ε is the number of standard deviations above or below the logarithmic mean. If ε is set to zero, the equation predicts median values of Y, which for a given scenario have a 50% probability of being exceeded. Setting ε equal to 1 gives the 84-percentile values of Y, a value of 2 yields the 97.7-percentile values and a value of 3 for ε will result in the 99.9-percentile values. The important point is that the equations predict a probabilistic distribution of values of Y rather than unique values as is often portrayed by plotting curves of the median values. The distribution of residuals is generally assumed to be log-normal, because it has been found that when the regression is performed on logarithms of the ground-motion values, the Gaussian distribution provides a very good approximation to the distribution of residuals.

Figure B-12 shows the range of predictions from a typical empirical GMPE bounded by ε values of -1 and +1, which means that for an earthquake of this magnitude, there is a 68% probability of the ground motions at a site at any particular distance falling within the shaded zone. The standard deviation (sigma, σ) is an indispensable part of the equation and can exert a very strong influence in PSHA as explained in Section B.3.3. This parameter is discussed in detail by Strasser et al. (2009) and Al Atik et al. (2010).

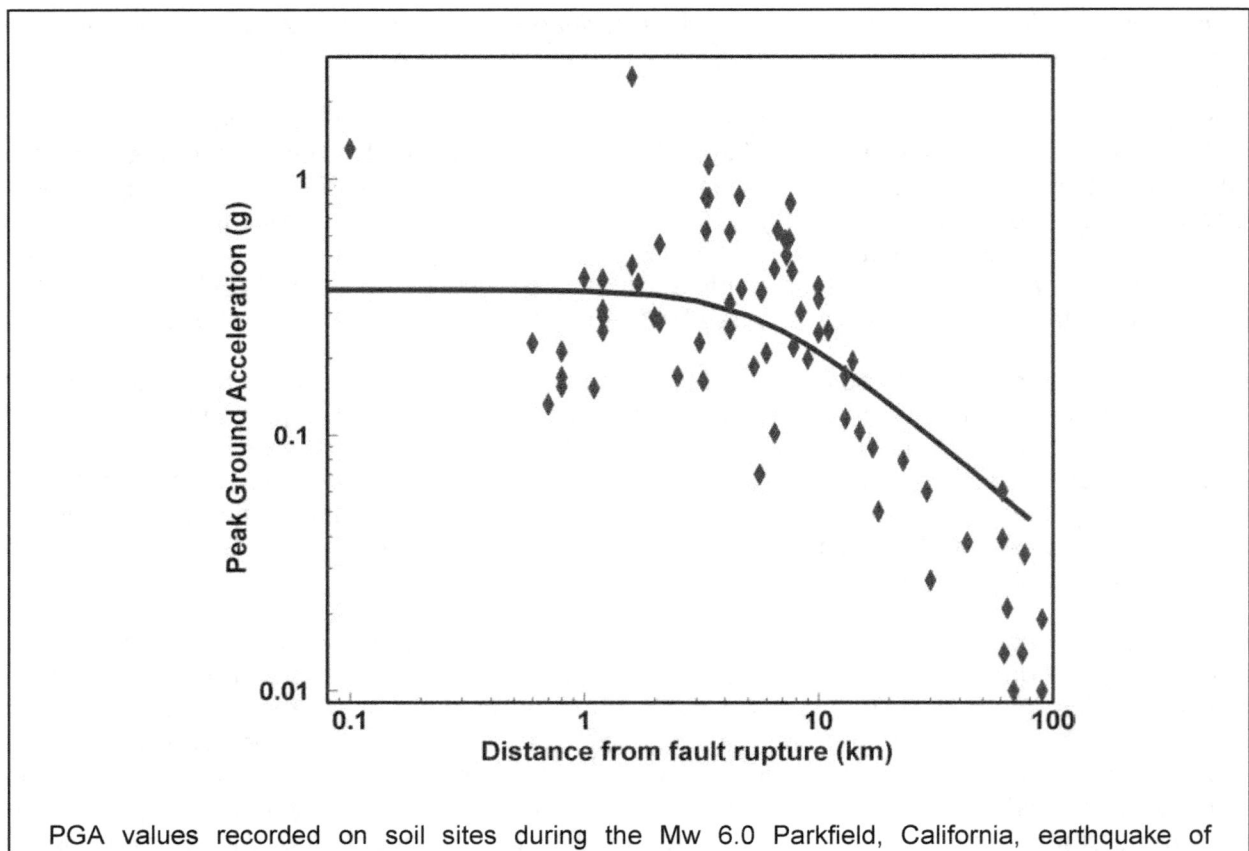

PGA values recorded on soil sites during the Mw 6.0 Parkfield, California, earthquake of September 2004, compared with the median predictions from the western North American empirical GMPE of Boore et al. (1997).

Figure B-11. PGA Values Recorded on Soil Sites Versus the Median Predictions from a Western GMPE.

In regions of low-to-moderate seismicity, the available strong-motion data are usually insufficient to allow the derivation of empirical equations through regression analysis. In such situations, models are generally derived primarily using stochastic simulations based on source, path and site parameters determined from the inversion of weak-motion recordings (Boore, 2003). Many models have been derived in this way for application to Eastern North America, such as Atkinson and Boore (2006). The variability in the ground-motion parameters obtained from multiple simulations is not sufficient to determine the value of σ to be associated with such models; an example of how appropriate estimates of variability may be determined for stochastic equations can be found in Toro *et al.* (1997).

In addition to the direct use of stochastic simulations to obtain predictions of ground motions in stable regions without indigenous strong-motion accelerograms, other approaches that have been proposed include the use of hybrid adjustments to models from other regions (Campbell, 2003), facilitated by inversions to obtain equivalent stochastic source, path and site parameters for empirical GMPEs from other regions (Scherbaum *et al.*, 2006). Another option is the referenced empirical approach in which local recordings of small-magnitude earthquakes are used to adjust a model from another region (Atkinson, 2008).

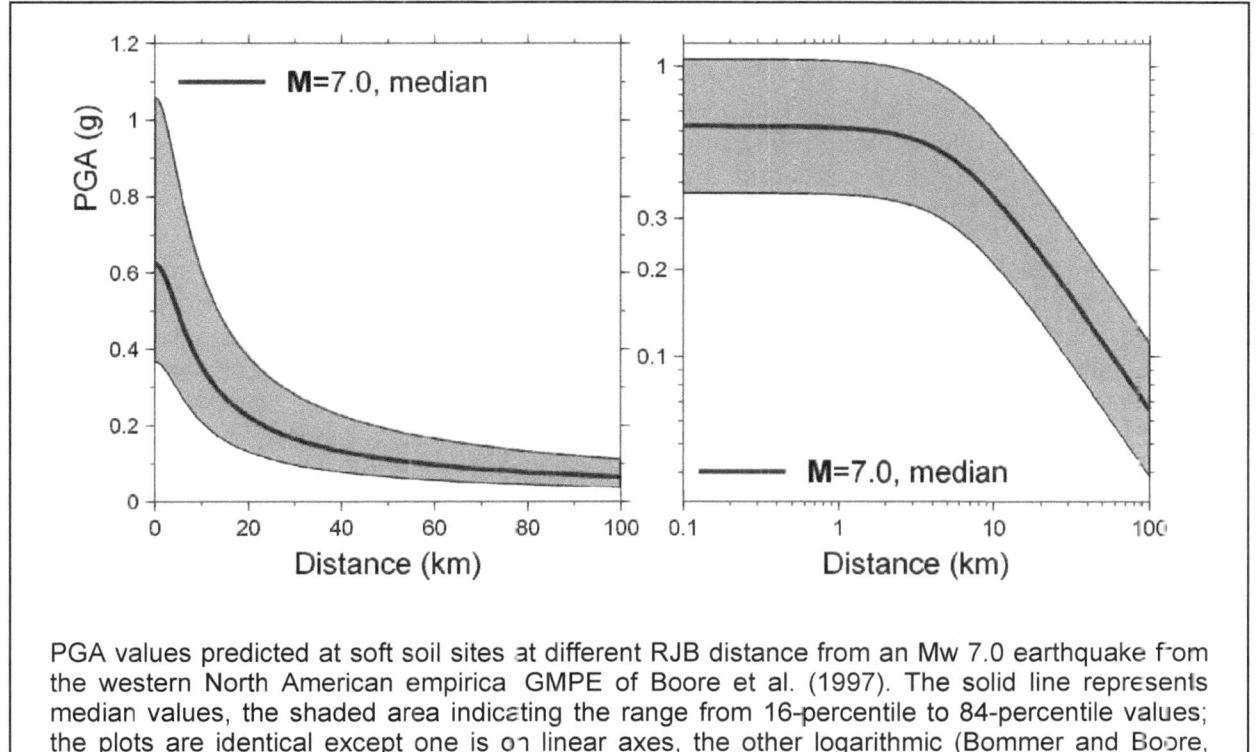

PGA values predicted at soft soil sites at different RJB distance from an Mw 7.0 earthquake from the western North American empirica GMPE of Boore et al. (1997). The solid line represents median values, the shaded area indicating the range from 16-percentile to 84-percentile values; the plots are identical except one is on linear axes, the other logarithmic (Bommer and Boore, 2004)

Figure B-12. PGA Values Predicted for Soft Soil Sites at Different Distances.

B.3.3 The Hazard Integral and PSHA Output

In deterministic seismic hazard analysis (DSHA), a limited number of earthquake scenarios are chosen and are defined in terms of magnitude and location (from which the distance is calculated). The ground motion is calculated at a chosen exceedance level (usually the median or 84-percentile value). By contrast, in PSHA all possible earthquake scenarios that could affect the site are considered, including all the possible exceedance levels of the ground-motion parameter for each combination of magnitude and distance, taking account of the rates of occurrence (and uncertainties) of each scenario. In practice, scenarios beyond certain distance are neglected, as are those smaller than the chosen values of m_{min} and the distribution of ground-motions is generally truncated at some level. The output from DSHA will be a single value of the ground-motion parameter for each scenario, and this will not automatically include an indication of its likelihood. PSHA yields exceedance rates for a wide range of values of the ground-motion parameter.

At this stage it is worth clarifying a point regarding terminology. In Section B.2 the word *hazard* was used to describe all of the potentially destructive effects of earthquakes, such as strong ground shaking, surface rupture, tsunamis and liquefaction. In the context of PSHA, by convention, the term *hazard* refers specifically to the rate (annual frequency) or probability of occurrence (within a specified period, usually one year) of particular values of each of those effects. This is potentially rather confusing, but when a probabilistic analysis is conducted in order to determine the hazard of ground shaking at a site, one should not refer to values of the ground motion, such as 0.3*g*, as being the hazard. Rather, the hazard is the rate at which 0.3*g* will be exceeded at the site of interest.

An overview of the history of PSHA is presented in McGuire (2008). The basic formulation of PSHA was presented by Cornell (1968), and subsequent refinements were presented by Esteva (1969, 1970), Cornell (1971), Merz and Cornell (1973) and Cornell and Merz (1975). The first widely-used code for computing probabilistic seismic hazard was developed by McGuire (1976).

In essence, PSHA is the process of integrating over all the possible earthquakes scenarios that may impact a site and taking account of both the scatter from the (random) aleatory processes described in the previous section and the likelihood that that particular scenario actually occurs. This assessment requires source characterization models for earthquake occurrence in space and time alluded to in the quote from Cornell (1968) at the beginning of Section B.3.2. It also includes integration over the distribution of full range of ground-motions predicted by the GMPEs. The later is required to fully take account of the uncertain level of ground shaking given a particular earthquake scenario. This second element was not included in the original formulation by Cornell (1968), possibly because the few GMPEs available at the time generally did not report sigma values. Although integration over the ground-motion variability was incorporated very soon afterwards, the practice of conducting seismic hazard analysis without this indispensable element persisted for a long time among some practitioners, leading to gross under-estimation of hazard (Bommer and Abrahamson, 2006).

There are several ways that the PSHA calculations can be understood, but essentially the process is the integration over the variables of magnitude (*m*), distance (*r*) and relative ground-motion exceedance level (ε). As an example, consider the source zone in Figure B-13 and a site just outside the source. Assume that the area source zone is now divided into elements of, say, 1 km x 1 km squares, and further assume, for simplicity, that the GMPE being used employs the r_{epi} distance metric. The recurrence relationship for each source area should be normalized by the area of the source so that the relationship predicts the actual number of earthquakes greater than or equal to a given magnitude per year per km^2 (averaged over as long a time period as possible). In the case of fault sources modeled as lines, the normalization will be per km of length, and for faults modeled as planes the normalization will be by area.

For this illustration, the hazard is being calculated for PGA, for which a series of target values will be considered. Assume for now that the first target is 0.2*g*. The contributions from every square will be considered in the calculations, up to a distance beyond which the effects of attenuation will be sufficient to make the contributions negligible. This distance will depend on the maximum magnitudes specified for the more distant sources, on the response frequencies of interest, and on the maximum value of ε considered in the analysis. A distance of approximately 300 km is likely to be sufficient in most cases but should be specified for each individual study.

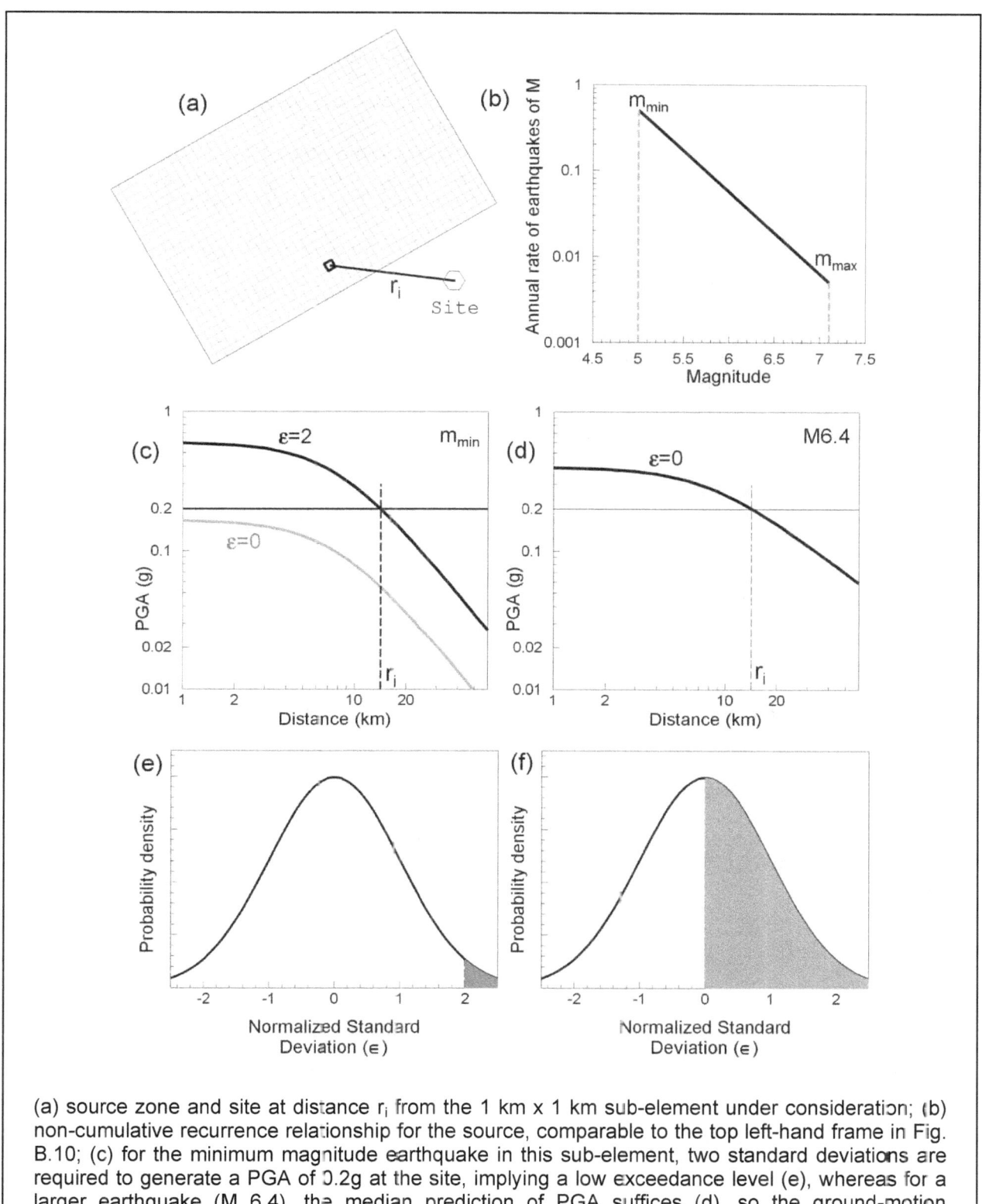

(a) source zone and site at distance r_i from the 1 km x 1 km sub-element under consideration; (b) non-cumulative recurrence relationship for the source, comparable to the top left-hand frame in Fig. B.10; (c) for the minimum magnitude earthquake in this sub-element, two standard deviations are required to generate a PGA of 0.2g at the site, implying a low exceedance level (e), whereas for a larger earthquake (M 6.4), the median prediction of PGA suffices (d), so the ground-motion exceedance is higher for this less frequent earthquakes scenario (f). See text for complete explanation.

Figure B-13. Illustration of the Mechanics of PSHA.

Taking one of the square elements, the distance r_j from the site to the center point of this 1 km x 1 km area is calculated. The range of magnitudes from m_{min} to m_{max} relevant to this seismic source is now divided into increments of Δm and each of these will be considered in turn. The annual rate of occurrence for the mid-range value, m_i, of each magnitude increment is calculated from the cumulative recurrence relationship in Equation (B.7) converted to its non-cumulative form. This conversion is made by dividing the magnitude range from m_{min} to m_{max} divided into steps of Δm, and the non-cumulative rate assigned to the central value of each magnitude bin is calculated from:

$$N(m_i) = n\left(m_i - \frac{\Delta m}{2}\right) - n\left(m_i + \frac{\Delta m}{2}\right)$$ Equation B.9

The value of the normalized residual ε_i that is required for the combination of m_i and r_j to yield a PGA of 0.2g at the site is then calculated.

The frequency with which this particular earthquake scenario will cause 0.2g to be exceeded at the site is the product of the annual rate of earthquakes of magnitude m_i and the exceedance probability of ε_i determined from the normal distribution. For m_{min}, which will have a high annual rate (0.5 in Fig. B-13b), a large value of ε_i may be needed to obtain 0.2g; if that value is 2, then the frequency of 0.2g being exceeded by this scenario will be 0.5 x 0.023 = 0.0115. Where 0.5 is the annual rate of earthquakes of magnitude m_i and 0.023 is the probability that ε=2 is exceeded (i.e., 1.00-0.977=0.023).

For a larger earthquake (magnitude 6.4, say) which will have much lower annual rate (say 0.02), the median value of PGA may be equal to the PGA of interest, in which case ε_i is zero and the annual exceedance frequency of 0.2g from this scenario is 0.02 x 0.5 = 0.010. The total frequency of exceedance of 0.2g at the site from earthquakes in this first 1 km x 1 km element is obtained by summing the frequencies of exceedance calculated in this way for all the magnitude increments. The total frequency of exceedance of 0.2g at the site is obtained by repeating these calculations for every elemental area over all of the seismic sources.

Mathematically, the process described in the previous paragraph can be expressed by the following equation to determine the annual rate at which the target PGA value, pga^*, is exceeded:

$$\lambda_{PGA}(pga^*) = \sum_i \left\{ \iiint I[PGA > pga^* | m,r,\varepsilon] \upsilon_i f_{M,R,E}(m,r,\varepsilon) \, dm \, dr \, d\varepsilon \right\}$$ Equation B.10

where $I(PGA>pga^*)$ is an indicator function taking a value of unity if PGA > pga^* and is zero otherwise; υ_i is the annual rate of exceedance of m_{min} in the seismic source, as defined in the relationship in Equation (B.7); the last term in the integrand is the probability density function of magnitude, distance and ground-motion exceedance (ε). Upper case letters in Equation (B.10) represent random variables, and lower case letters represent realizations of these variables. The limits on the integrals have been discussed earlier in this section for distance and in the previous section for magnitude. The third variable, ε, is related to the normal distribution and therefore has no natural upper or lower limit, although clearly small negative values of ε will have no influence on the hazard results by virtue of producing extremely small accelerations at the site. Many hazard analysts have imposed upper limits on the value of ε at 2 or 3, which can have a pronounced effect in reducing the calculated hazard. However, the currently available datasets do not provide any justification for truncating the ground-motion distribution at these low levels (Strasser *et al.*, 2008).

The earthquake recurrence rate, v_i, can be taken outside the integral, and the equation can be re-written as follows:

$$\lambda_{PGA}(pga^*) = \sum_i \left\{ v_i \int_{r_{min}}^{r_{max}} \int_{m_{min}}^{m_{max}} P\left[PGA > pga^* | m,r\right] f_{M,R}(m,r) \, dm \, dr \right\}$$

Equation B.11

where the first term in the integrand is simply the probability of exceeding the target acceleration for a given magnitude-distance combination. This probability is simply a function of the value of ε, which is the residual normalized by the standard deviation:

$$\varepsilon^* = \frac{ln(pga^*) - \mu_{PGA|M,R}}{\sigma_{ln(PGA|M,R)}}$$

Equation B.12

Then the probability indicated by the first term in the integrand in Equation (B.11) is:

$$P[PGA > pga = P[\varepsilon > \varepsilon^*] = 1 - \Phi(\varepsilon^*)$$

Equation B.13

where $\Phi(x)$ is the standard normal cumulative probability density function.

The calculations described above (for 0.2*g*) can then be repeated for a range of values of PGA in order to obtain a hazard curve that provides the annual frequency of exceedance for each value of PGA. One thing that may strike the reader at this point is that although the method being discussed here is known as probabilistic seismic hazard analysis, the calculations presented have focused on annual rates of exceedance of particular ground motion levels, rather than on their probabilities. The annual probability of exceedance can be calculated if a probability distribution is adopted to represent the temporal behavior of seismicity, the most commonly used being the Poisson distribution. The annual probability of exceeding the target acceleration value can then be calculated from the simple equation:

$$P(PGA > pga^*) = 1 - e^{-\lambda}$$

Equation B.14

where λ is defined in Equation (B.11). For small values of λ, the annual probability and the annual rate of exceedance are almost identical (Figure B-14), which leads to their often being used interchangeably.

A term that was widely used in the past is the "return period", which is simply the reciprocal of the annual rate of exceedance (*i.e.*, 1/ λ), and expressed in units of years. The return period of a level of ground motion should not be confused with the recurrence interval (or repeat time) of an earthquake of a particular magnitude or greater, which is the reciprocal of $N(m)$ in Equation (B.7). These two terms are also often used interchangeably, which is no longer correct because by established convention they now refer to different quantities. In addition, the use of the term "return period" to refer to hazard results generates significant misinterpretation of hazards results by non-practitioners.

Hazard calculations can be conducted for any ground-motion parameter for which a prediction equation is available. In Figure B-15 there are hazard curves for PGA and for spectral accelerations at response periods of 0.2 and 1.0 second. By calculating the hazard curves for spectral accelerations at a large number of response periods (or inversely response frequencies) and then reading off the ordinates at a selected annual exceedance probability, uniform hazard spectra (UHS) can be constructed (Figure B-16).

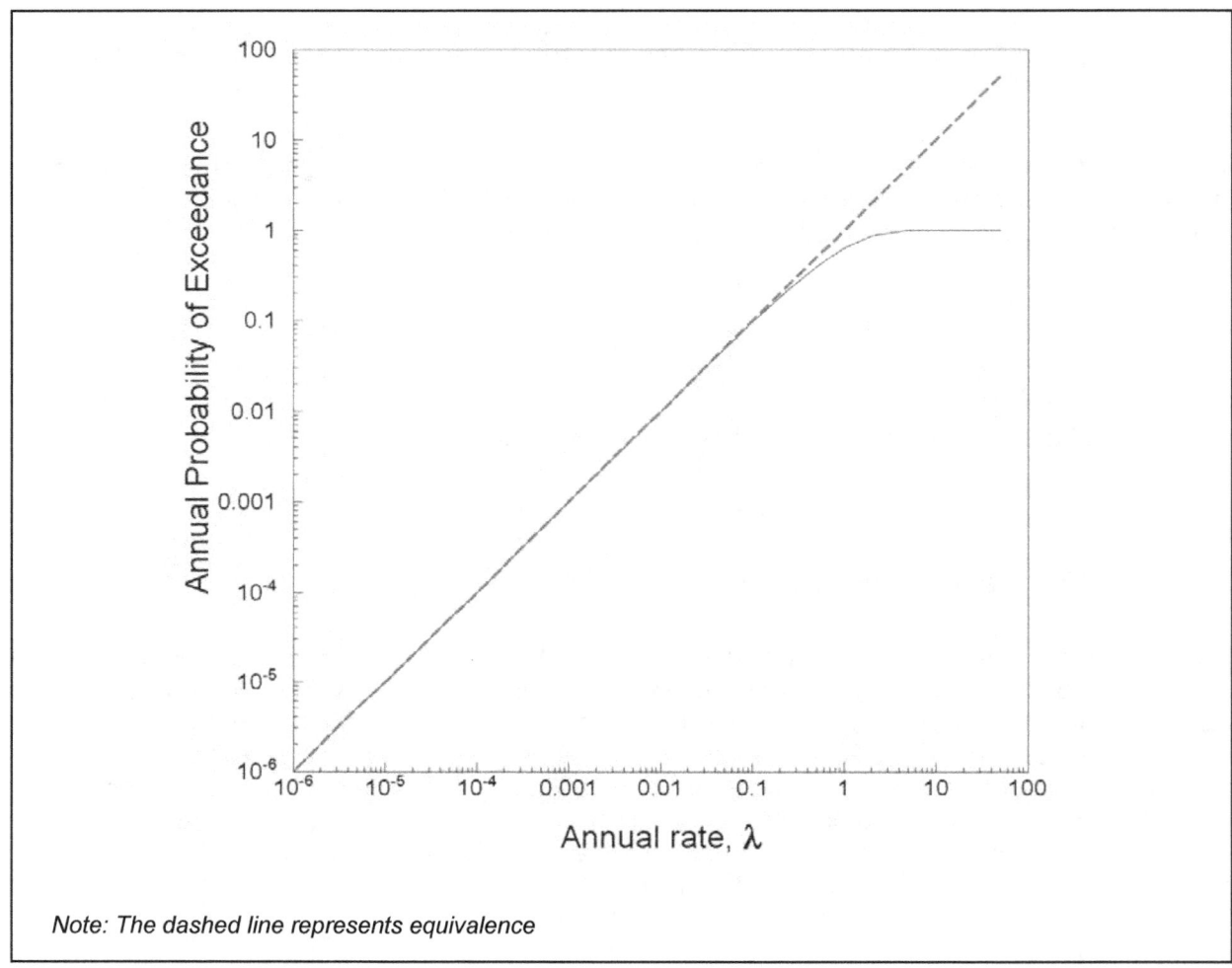

Note: The dashed line represents equivalence

**Figure B-14. Relationship Between Annual Rate of Exceedance and
Annual Probability of Exceedance for a Poisson Process.**

The UHS represents the maximum spectral acceleration that is expected to be exceeded at the site for a range of response periods for a specified annual rate or probability of exceedance. A response spectrum obtained by anchoring a standard response spectral shape to a value of PGA (or some other spectral frequency) determined through PSHA is unlikely to have the same frequency of exceedance at all response periods, and hence will not be of uniform hazard (McGuire, 1977).

The UHS is a useful representation of the site hazard. If the only quantity of interest is the response of the fundamental mode of vibration of the structure, the UHS is all that is required. However, for more advanced structural analyses, such as those using modal superposition techniques or dynamic analysis requiring acceleration time-series as input. it is not an appropriate representation. For such applications, it is desirable to identify individual earthquake scenarios, which can be achieved through the process of disaggregation[22] (McGuire, 1995; Bazzurro and Cornell, 1999). Because PSHA is a process of integrating the contributions of all possible earthquake scenarios in order to obtain annual exceedance frequency of a given ground motion, disaggregation is really little more than a book-keeping exercise that groups the contributing scenarios into bins of magnitude and distance. For a chosen ground-motion parameter and exceedance frequency, disaggregation can identify the relative contributions of different magnitude-distance bins to the total hazard. The contributions from different intervals of ε can also be shown (Figure B-17).

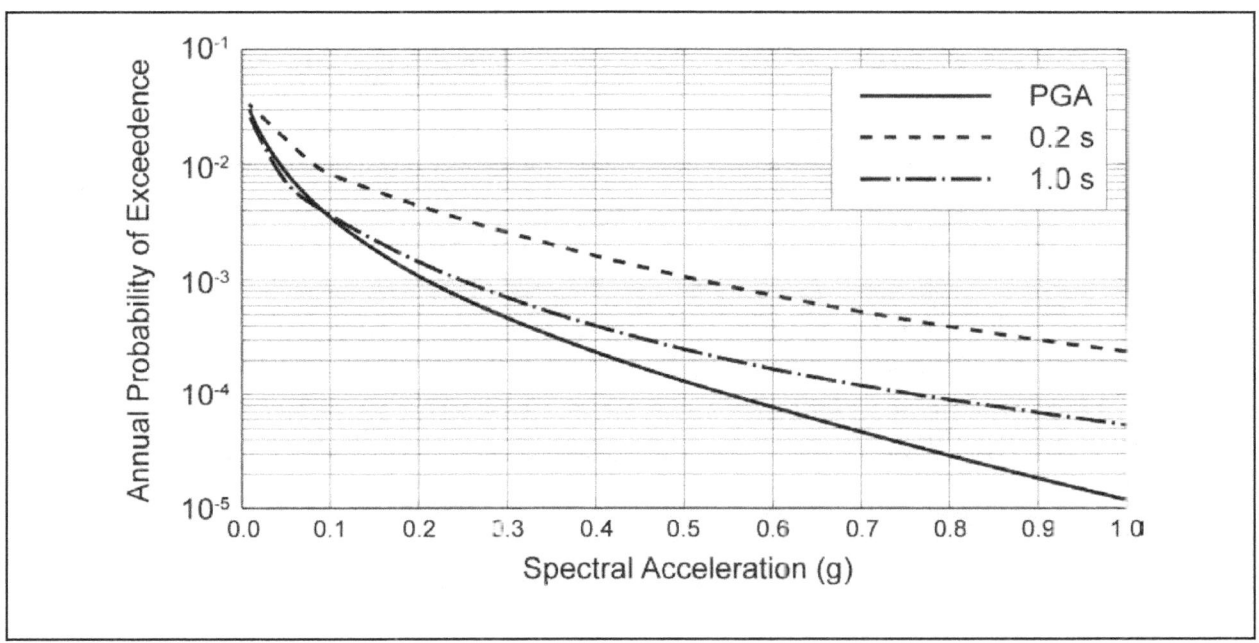

Figure B-15. Examples of Seismic Hazard Curves for a Site in Terms of Spectral Accelerations at Different Response Periods (ASCE, 2004)

---

[22] The term de-aggregation is also widely used.

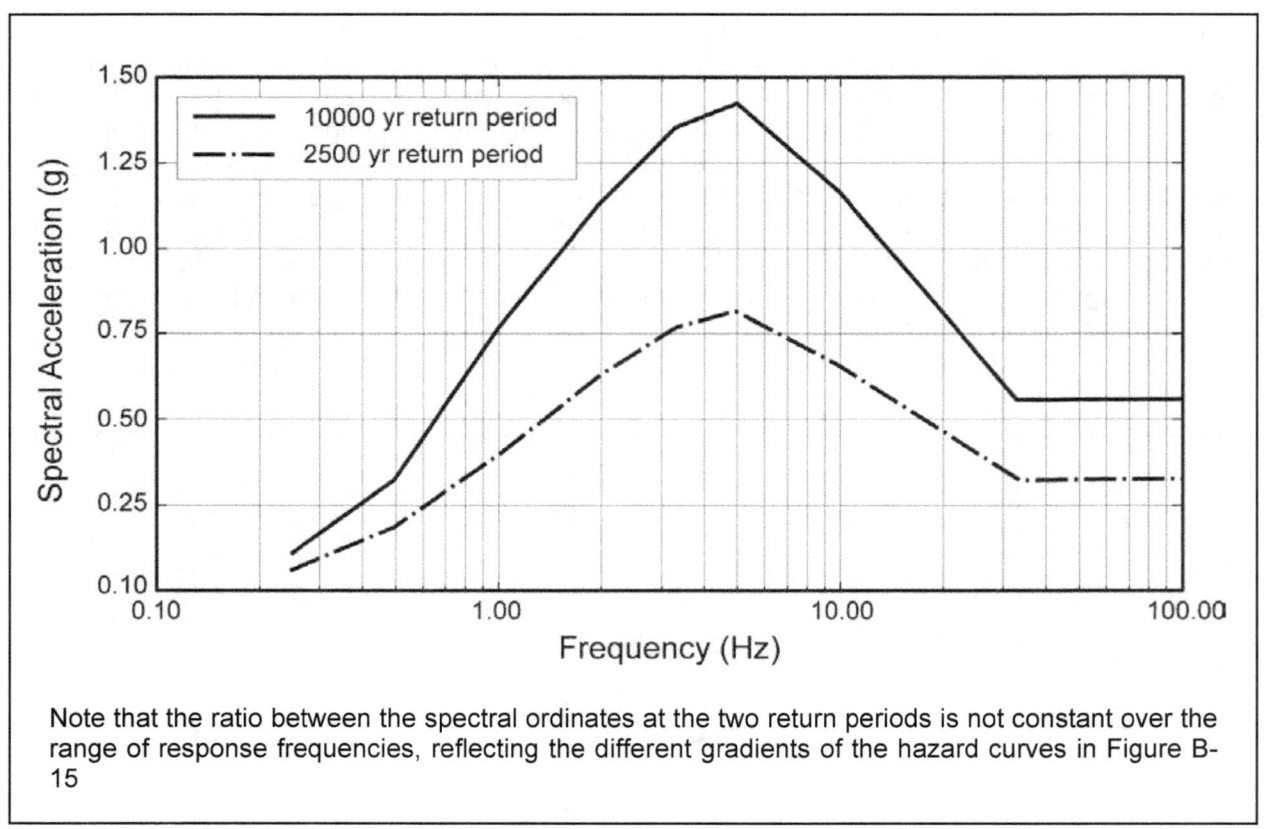

Note that the ratio between the spectral ordinates at the two return periods is not constant over the range of response frequencies, reflecting the different gradients of the hazard curves in Figure B-15

Figure B-16.   Examples of Uniform Hazard Response Spectra (ASCE, 2004).

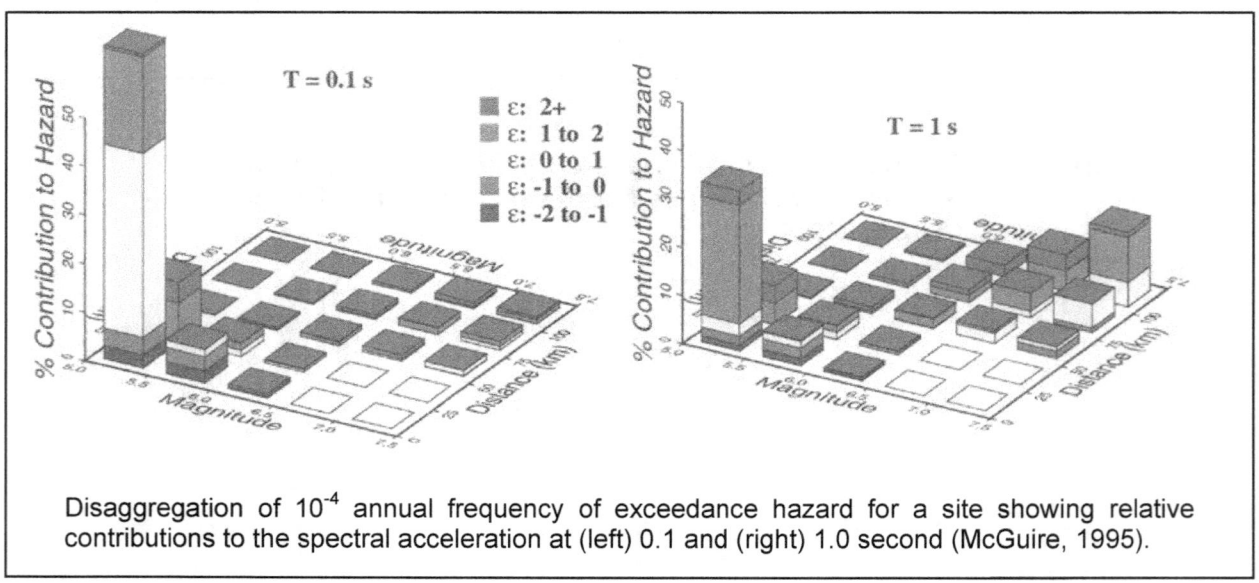

Disaggregation of 10^{-4} annual frequency of exceedance hazard for a site showing relative contributions to the spectral acceleration at (left) 0.1 and (right) 1.0 second (McGuire, 1995).

Figure B-17.   Example of Disaggregation.

B.3.4 Epistemic Uncertainty

In the previous section, the execution of PSHA calculations was described for a single seismic source characterization model and a single ground-motion prediction model, resulting in a unique hazard curve for each ground-motion parameter considered. This would be an acceptable approach if there were no epistemic uncertainty and the analyst could select a single model for each source and process and a single value for each parameter with confidence. In practice, there is always considerable uncertainty associated with most of the inputs, which is reflected in the fact that different experts are likely to select different models or parameter values as being the most appropriate to a given application. The SSHAC guidelines make the following statement in this regard:

> "Despite much recent research, major gaps exist in our understanding of the mechanisms that cause earthquakes and of the processes that govern how an earthquake's energy propagates from its origin beneath the earth's surface to various points near and far on the surface. The limited information that does exist can be – and often is – legitimately interpreted quite differently by different experts, and these differences of interpretation translate into important uncertainties in the numerical results from a PSHA" (NUREG/CR-6372).

Uncertainties will be encountered in every input to PSHA calculations. For example, in terms of seismic source characterization, for diffuse seismicity there will be the choice as to whether area source zones or smoothing kernels are the more appropriate representation. For either option, there will then be choices to be made during the process of source characterization. One must determine the limits of the source zones (Figure B-18) in the first approach, or in the dimensions of the smoothing kernel in the second approach.

As noted previously, it will not always be unambiguously clear whether a Gutenberg-Richter or characteristic earthquake recurrence model should be adopted for fault sources. When deriving the recurrence relationships from seismicity data, assumptions need to be made about completeness of the earthquake catalog as a function of time, location, and magnitude. Another obvious example of epistemic uncertainty in seismic source characterization is the selection of appropriate values of the limiting maximum magnitude, m_{max}, in each source.

In terms of ground-motion modeling for a PSHA, the fundamental epistemic uncertainty is the choice of GMPE for the prediction of the median ground motions. Even in regions with abundant strong-motion data, analysts may make different decisions about which parts of the data should be used in the regressions and which functional form should be used. Figure B-19 shows predicted median values of PGA on rock sites from the NGA equations derived for application in western North America. This figure illustrates the nature of epistemic uncertainty in median ground-motion predictions. The predicted values are generally in good agreement, but there are differences despite a common database (Chiou et al., 2008).

The differences are smallest for magnitudes (M_w) 6 and 7, where the data are most abundant (Figure B-20). At magnitude 5 and magnitude 8 there is greater divergence among the model predictions. These curves are for strike-slip earthquakes, which is the case for which the models are in closest agreement. For normal- and reverse-faulting earthquakes, there is greater spread among the predictions.

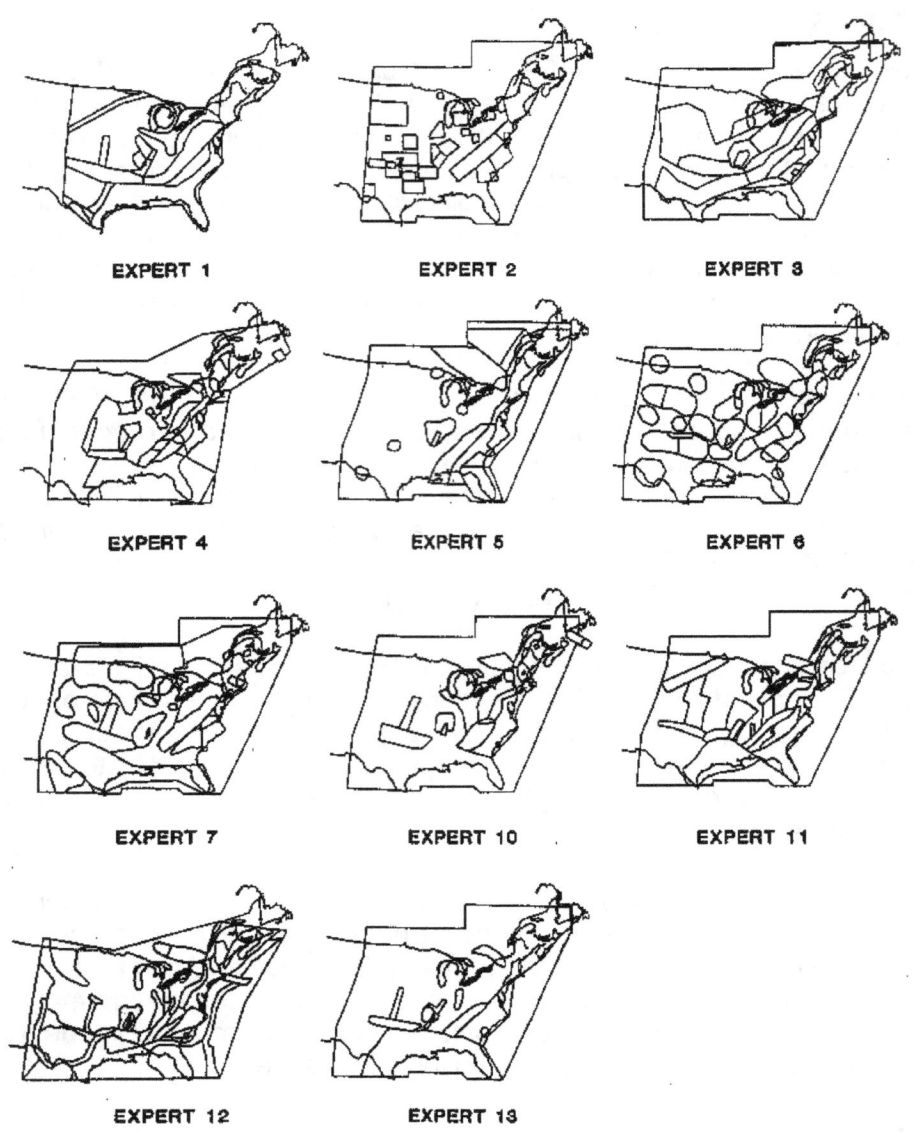

EXPERT 1 EXPERT 2 EXPERT 3

EXPERT 4 EXPERT 5 EXPERT 6

EXPERT 7 EXPERT 10 EXPERT 11

EXPERT 12 EXPERT 13

Figure B-18. Different Source Zone Geometries Defined by Different Experts for PSHA at Sites in the Central and Eastern United States (Bernreuter et al., 1989).

When assessing the epistemic uncertainty in median ground-motion predictions, consideration should be given to the fact that strong-motion datasets are generally sparse, particularly at larger magnitudes, and may be biased. For example, a database may contain recordings from a small number of large-magnitude events, which all happen to have had above (or below) average stress drops for the region in question. In the case of the NGA models, this consideration is relevant also because all of the events of magnitude M_w 7.5 and greater represented in the database occurred outside the United States, and the implicit assumption is made that future California earthquakes of this size will have similar characteristics. Worthy of note in this respect is the fact that in deriving the current national hazard maps for the United States, the USGS used the NGA models for California but added additional models with higher and lower median predictions for magnitude ranges in which the NGA database is relatively sparse (Petersen *et al.*, 2008).

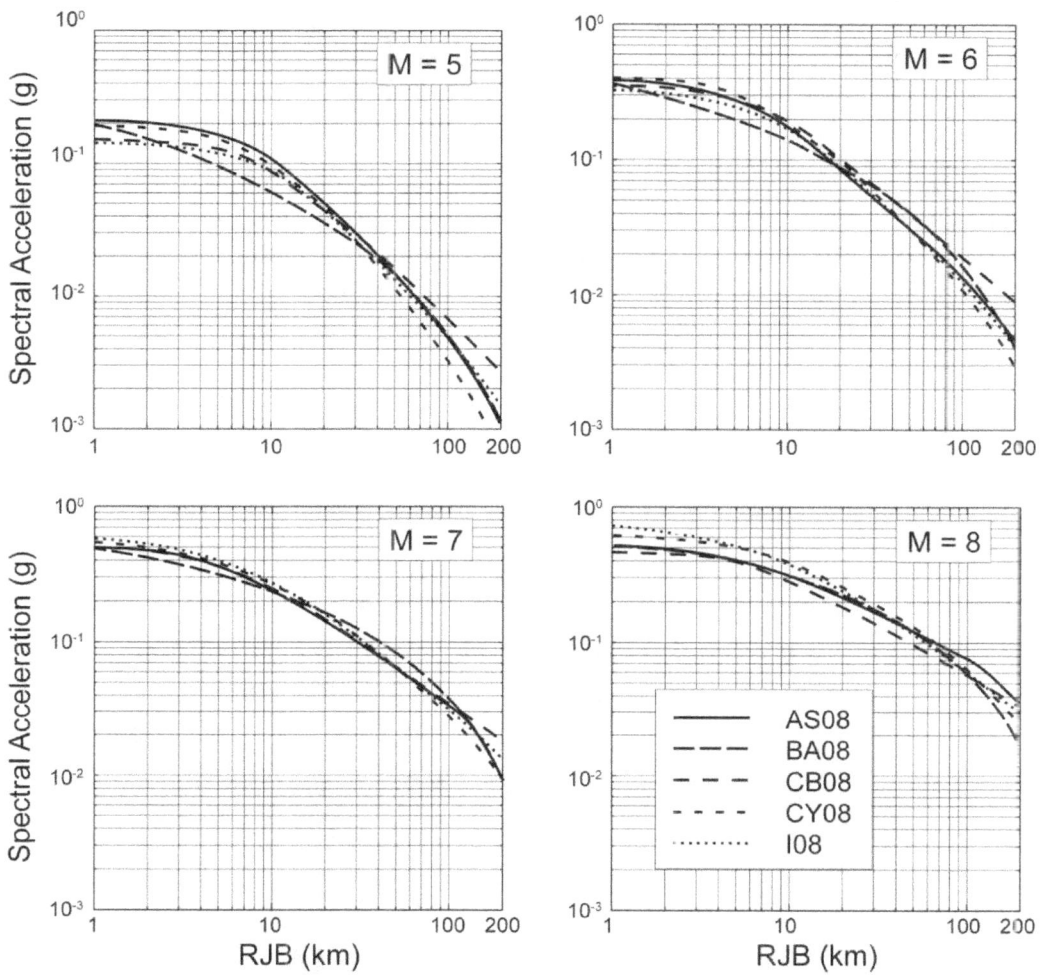

Figure B-19. Comparisons of Predicted Median PGA Values from the NGA Models for Rock Sites and Strike-Slip Earthquakes of Different Magnitudes (Abrahamson *et al.*, 2008).

In regions of low-to-moderate seismicity (e.g., the Central and Eastern United States) where earthquakes are less frequent, epistemic uncertainty in the hazard inputs will inevitably be larger because the available databases are sparser. This applies equally to the seismic source models and the ground-motion prediction models. For the latter, the epistemic uncertainty will generally be particularly large for the upper range of magnitudes considered in the hazard calculations because it is likely that there will be few if any available recordings from regional earthquakes of such size. Consideration should be given not only to the epistemic uncertainty as reflected in the range of legitimate interpretations of observed events, but also to the greater uncertainty associated with projections to future events that may lie beyond the existing range of observations.

In addition to the epistemic uncertainty associated with the median ground-motion predictions, there may be epistemic uncertainty in the associated value of standard deviation (i.e., the sigma). This may arise because the standard deviation can be modeled as a constant value for a given ground-motion parameter (homosceadastic variability) or as varying with one or more of the explanatory variables in the model (heteroscedastic variability). For example, among the NGA models, two have constant sigma values whereas three found sigma to be dependent on magnitude (Figure B-21).

Most of this report is about identifying and quantifying epistemic uncertainty in the inputs to the assessment of seismic hazard (or any other geo-hazard). Chapters 3, 4 and 5 of the report are focused on the use of multiple expert assessments to quantify uncertain inputs to hazard analysis. In the next section, the tools that are used for displaying and handling epistemic uncertainty in hazard analysis are briefly introduced.

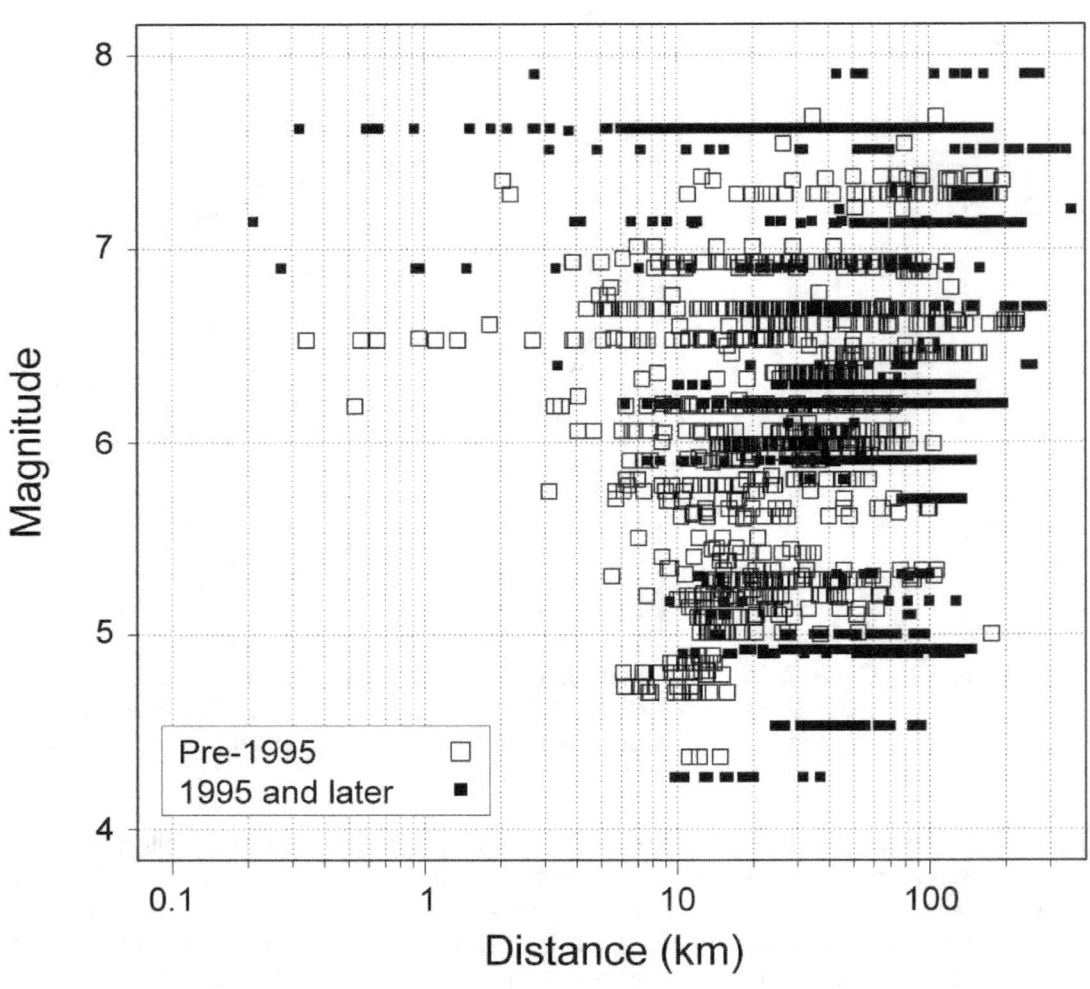

Figure B-20.   Magnitude-Distance Distribution of the NGA database (Chiou et al., 2008).

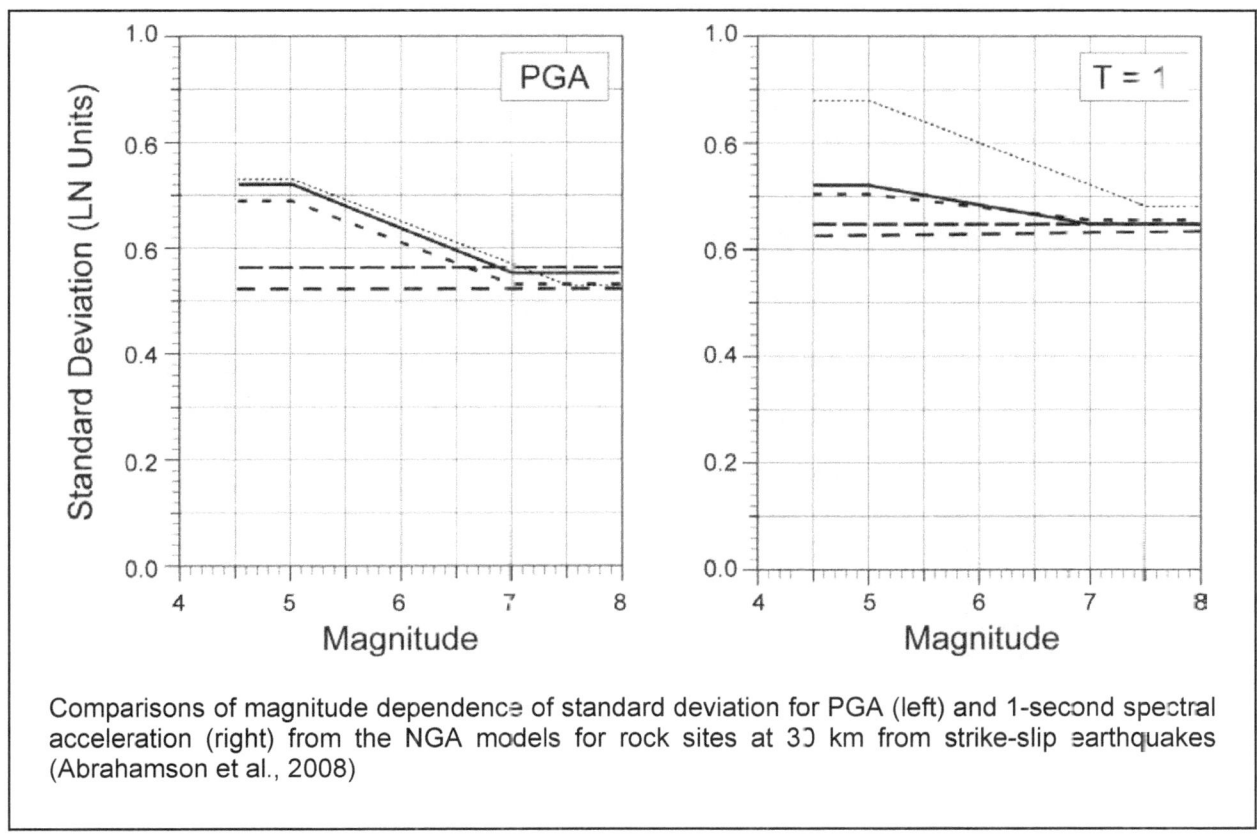

Comparisons of magnitude dependence of standard deviation for PGA (left) and 1-second spectral acceleration (right) from the NGA models for rock sites at 30 km from strike-slip earthquakes (Abrahamson et al., 2008)

Figure B-21. Comparisons of Magnitude Dependence of Standard Deviation.

B.3.5 Logic-Trees and Uncertainty in Hazard

The most widely used tool to represent alternative models for input to PSHA is the logic-tree (Kulkarni *et al.*, 1984; Coppersmith and Youngs, 1986; Reiter, 1990). The concept of the logic-tree is very simple: for each input, branches are set up for alternative models or alternative parameter values, and weights are assigned to each branch to reflect the relative confidence that the analyst has in each model being the best representation of that component of the hazard input. The weights on branches emanating from a single node on the logic-tree are assigned to sum to unity (Figure B-22) because they are subsequently used as probabilities.

The hazard calculations are then performed using every possible combination of branches. Each individual combination of branches has an associated individual hazard estimate and an associated total probability that is determined by calculating the product of all the branch weights involved in that calculation. For any given value of the selected ground-motion parameter, the final product is a suite of annual exceedance frequencies (hazard estimates) and their associated probabilities. The statistics of the hazard estimates can then be easily calculated, yielding fractiles and the mean hazard (Figure B-23).

Whereas the aleatory variability influences the shape (gradient and curvature) of an individual hazard curve, the epistemic uncertainty gives rise to a suite of hazard curves. In risk analyses, all of the fractiles of the hazard may be used, but design of nuclear power plants is usually based on the ground motions associated with the mean hazard.

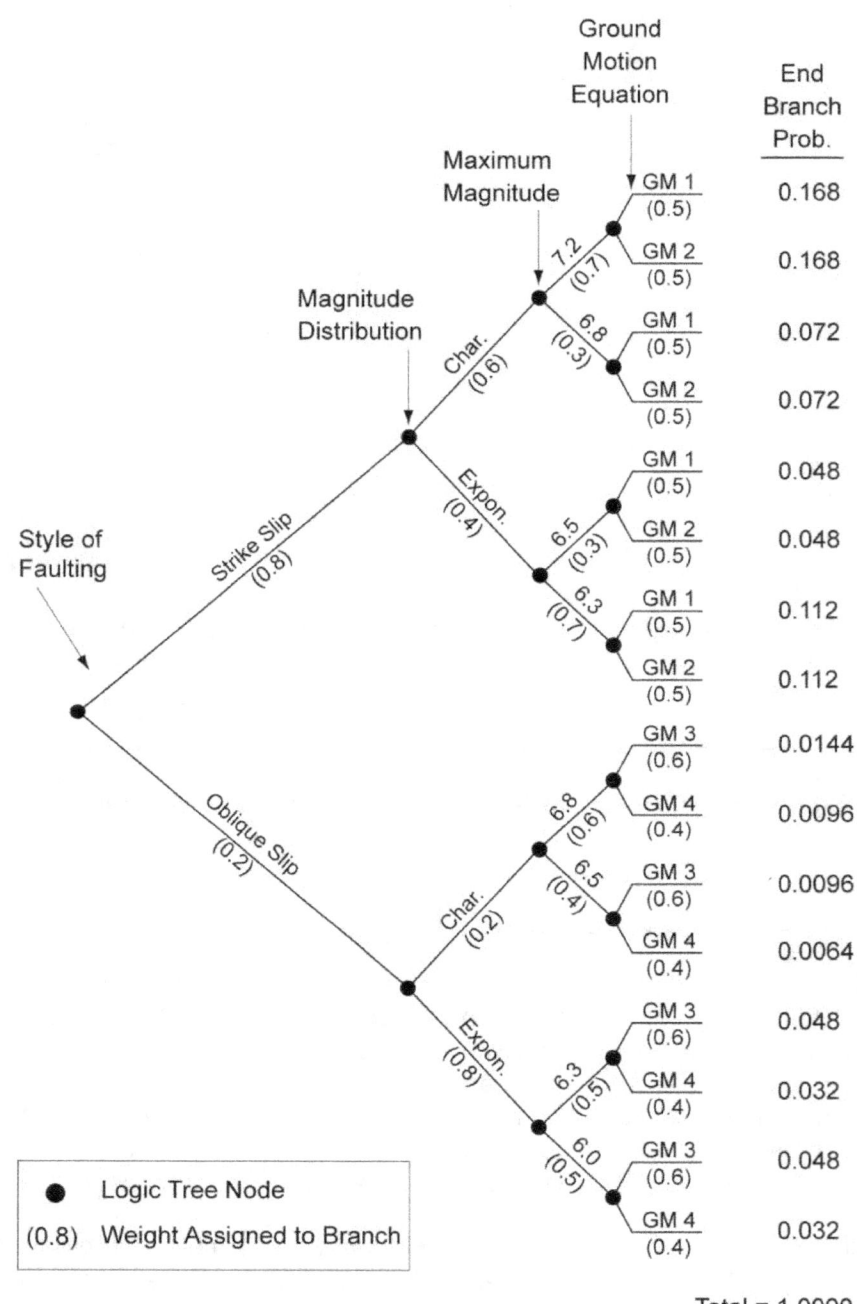

Figure B-22. Logic-Tree for a Single Fault Source (McGuire, 2004).

Although the logic-tree is an apparently simple tool with significant benefits that include transparency of the model, its use in PSHA can be conceptually problematic. In calculating the mean hazard, the branch weights are treated as (subjective) probabilities. The conceptual consequences of treating the weights as probabilities are discussed by Bommer and Scherbaum (2008). In calculating statistics of the hazard, such as the mean value, the weights on the logic-tree branches are treated as probabilities. This implies that the different models on the branches emerging from an individual node are mutually exclusive and collectively exhaustive (Bommer and Scherbaum, 2008). In practice, meeting these so-called MECE criteria can be very difficult, particularly in terms of the models being mutually exclusive given that the datasets from which different ground-motion prediction equations are derived, for example, will often overlap by including common records. The evaluators can, however, take this feature into account when assigning weights. The evaluators should put great emphasis on the selection of models to populate the branches of the logic-tree because once there are more than three or four models the mean hazard results tend to become rather insensitive to the actual weights assigned in the logic tree (e.g., Sabetta et al., 2005; Scherbaum et al., 2005).

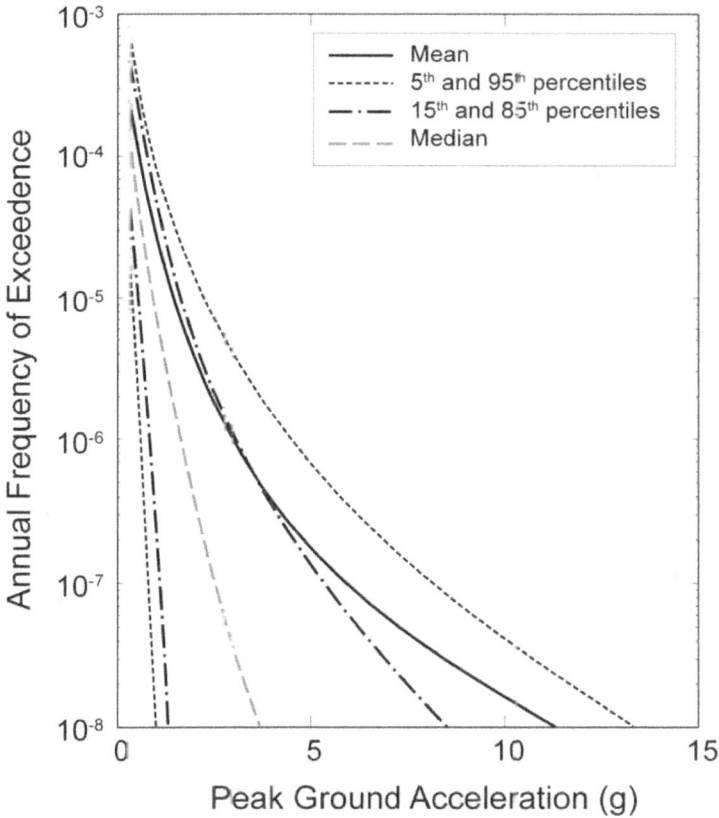

Figure B-23. Fractiles and Mean Hazard Curve for Rock Outcrop Location at Yucca Mountain (Abrahamson and Bommer, 2005).

Logic-trees are almost ubiquitous in PSHA nowadays but they are sometimes misused. Logic-tree nodes are reserved exclusively for alternative models or parameters (epistemic uncertainty) and not for the relative frequency with which things can happen. For example, the distribution of future earthquake focal depths in a region is an example of aleatory variability yet it is common to see a depth distribution approximated by branches on a logic-tree (Bommer and Scherbaum, 2008). Whereas it would be legitimate to include branches representing alternative models of focal depth distributions, the distributions themselves should be included in the hazard integrations because the logic-tree branches should only include epistemic uncertainties. Another common pitfall is to assign the a- and b-values of recurrence relationships to separate consecutive nodes, from which "*unintentional combinations of a- and b-values can result if the correlations between a and b are not defined*" (NUREG/CR-6372). If m_{max} is also added as a third uncorrelated node, then some combination of implied moment rate may be far outside the limits compatible with geodetic and geological data (Bommer and Scherbaum, 2008). Abrahamson (2000) gives additional examples related to incorrect formulations with regards to segmented *vs.* non-segmented fault rupture, which violate the criterion of mutually exclusivity.

For the ground-motion section of the logic-tree, a key issue is to ensure compatibility among the definitions employed for both the predicted ground-motion parameter and all the explanatory variables in the models (e.g., magnitude scale, distance metric, and site classification). Empirical relationships between different parameter definitions can be applied to adjust the parameters to a common convention, but the variability associated with these adjustments must be propagated to the sigma value of the equation (Bommer *et al.*, 2005).

Logic-tree branches are not the only option for representing alternative inputs to PSHA, although they do represent a very convenient tool for representing alternative models and their dependencies. As noted by McGuire (1993), there are essentially two ways that epistemic uncertainty can be represented:

> "*It is the case for all inputs to the seismic hazard analysis that alternative interpretations must be made where significant uncertainty exists. Thus the analyst should make a 'best' interpretation and should also represent uncertainties caused by lack of data or lack of knowledge either with a specified distribution or with alternatives.*"

Where the branches represent alternative approaches, such as area sources *vs.* smoothed seismicity for example, or discrete alternative conceptual models, it is clear that continuous distributions cannot be used. The choice between branches (discrete distributions) and continuous distributions for other parameters and models needs to be considered in terms of the relative merits and disadvantages of each approach in each particular application, bearing in mind that continuous distributions can be represented by discrete multiple-point approximations (e.g., Keefer and Bodily, 1983, and references therein).

There are many cases where a distribution is the natural choice, such as continuous parameter values. One such example is the maximum magnitude, m_{max}, within a seismic source. There is one true value of m_{max}, but it is not known what that single value is. Therefore, the analyst may specify branches with values of, say, 6.8, 7.0, and 7.3, with weights of 0.2, 0.6 and 0.2, but this clearly means that the maximum magnitude could take any value between 6.8 and 7.3, and not that it can only take one of these three values. In such situations, branches can be used but a distribution is probably more appropriate. With regards to this issue, the SSHAC guidelines stated "*If continuous probability distributions are discretized to apply the complete enumeration method, the proper choice of representative values and probabilities is important to derive an adequate estimate of the probability distribution for seismic hazard*" (NUREG/CR-6372).

For the case of ground-motion prediction models, the choice is not straightforward. In the original SSHAC guidelines it was envisaged that the epistemic uncertainty would be represented by distributions on medians and on standard deviations, characterized by σ_μ and σ_σ (NUREG/CR-6372). This approach was adopted for the PSHA conducted for Yucca Mountain (Stepp *et al.*, 2001). In contrast, for the PEGASOS project in Switzerland (Abrahamson *et al.*, 2002) it was decided to construct the ground-motion logic-tree with weights assigned to models, with separate branches for the medians and the sigmas (Figure B-24).

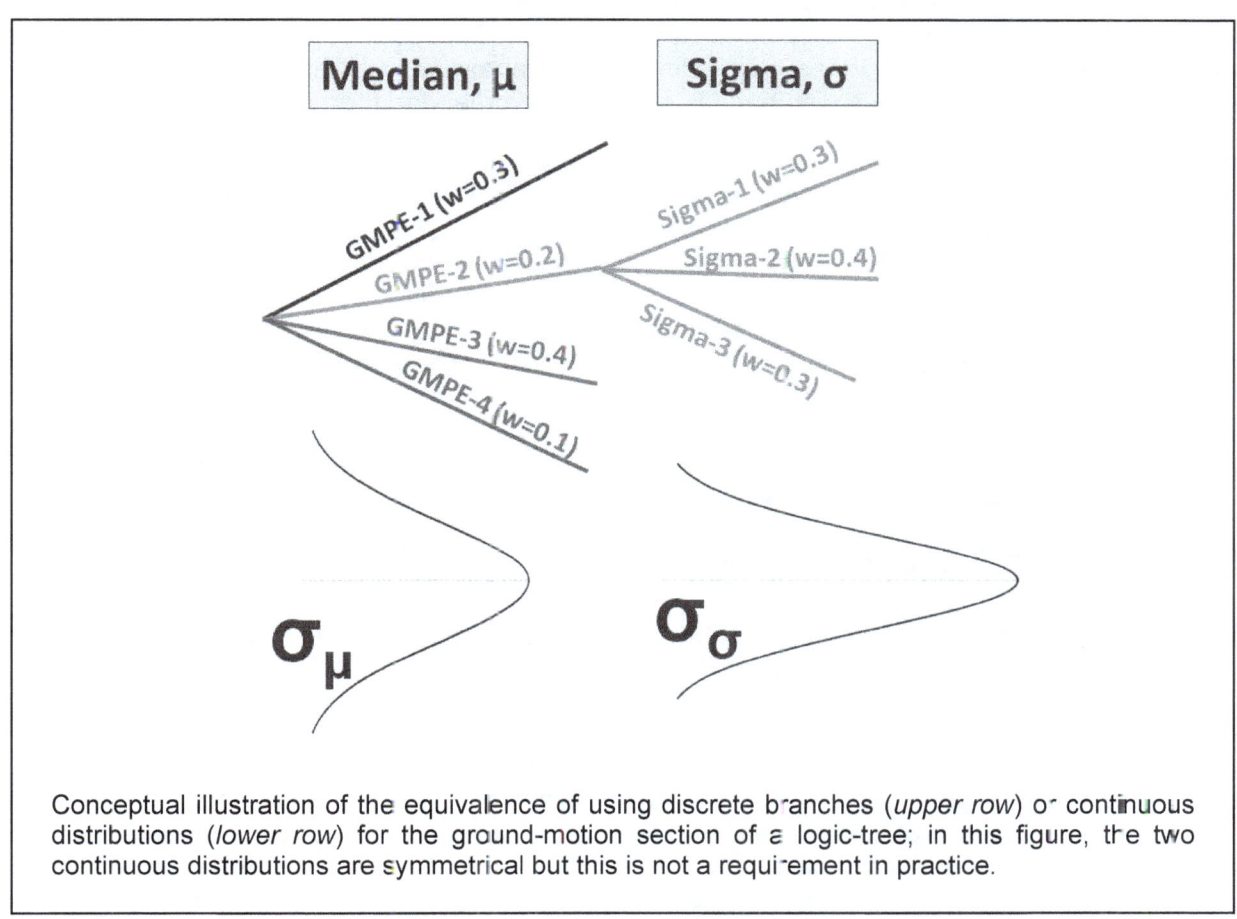

Conceptual illustration of the equivalence of using discrete branches (*upper row*) or continuous distributions (*lower row*) for the ground-motion section of a logic-tree; in this figure, the two continuous distributions are symmetrical but this is not a requirement in practice.

Figure B-24. Conceptual Illustration of the Equivalence of Using Discrete Branches or Continuous Distributions for the Ground-Motion Section of a Logic-Tree.

This decision was made because t was easier to implement and the resulting logic-tree structure was more transparent (Hanks *et al.*, 2009). However, this resulted in the experts effectively assigning weights to models rather than to ground-motion values (e.g., Scherbaum *et al.*, 2005). Because ultimately it is ground-motion values that are of interest, the distributions approach[23] used in the Yucca Mountain study, although more cumbersome, offers the advantage of a more direct relationship. An analyst assigning equal weights to selected ground-motion models, on the basis of having no criteria to rate one as being more favorable than another, is unlikely to be assigning equal weights to median ground motion or to measures of

---

[23] as opposed to the approach of discrete models on branches.

ground-motion variability due to the non-linear relationship. However, if the expert *does* have a reason to chose one ground-motion model over the other, this approach is appropriate. One potential disadvantage of adopting distributions rather than discrete models for ground motions is the tendency in the former case for experts to assign symmetrical distributions, which is less likely to occur (at least, it is less likely to intentionally occur) when using branches.

The preceding discussion aims only to highlight the issues under consideration when making the choice between these two options, and no suggestion is made in favor of selecting one approach over the other. The only conclusion from this discussion is that with respect to the original SSHAC Guidelines, those conducting hazard analyses should not feel constrained to adopting the use of continuous distributions as was recommended therein.

In closing this discussion, it is worth reiterating that uncertainty exists and needs to be captured. Therefore, an approach that captures uncertainty more effectively should be considered desirable from a nuclear safety perspective. As McGuire (1993) states very succinctly, "*The large uncertainties in seismic hazard are not a defect of the method. They result from lack of knowledge about earthquake causes, characteristics, and ground motions. The seismic hazard only reports the effects of these uncertainties, it does not create or expand them.*" For this reason, it is important not to refer to any analysis or approach 'giving higher hazard'. The true hazard at any location is a characteristic of Nature, and all any analysis method can give is an estimate of this quantity. The hazard exists and the analysis method chosen does not increase or reduce it; but appropriate approaches increase confidence that the hazard is being well characterized along with the associated uncertainties to the extent possible.

B.4   Testing Seismic Hazard Studies

PSHA is an attempt to characterize Nature, specifically in terms of the expected rate at which different levels of ground motion will be exceeded at a particular site. Faced with so many uncertainties associated with the inputs to a hazard analysis, it is natural for the analyst to wish to find corroboration for the outcome. In other words, the hazard evaluators will wish to ask the question: do the hazard estimates make sense? The term 'sanity check' is often applied to the attempts to address this question. This term (which is probably a misnomer in all cases) is not helpful because it does not clearly identify exactly what is being tested or explored in the hazard model or results. Clarity with regard to the goal and technical basis of an exploratory study is vitally important because it focuses attention on what can be expected and what the implications of the outcome might be in terms of increasing or undermining confidence in the hazard assessment.

A discussion of testing seismic hazard analyses is included here rather than in the main body of this report, because it is not an integral or indispensable part of a SSHAC Level 3 or 4 study. Every possible effort should be taken in PSHA studies to check for errors: independent checks on sample calculations and reproduction of known or published results using models implemented in the study are an important part of the process. Sensitivity studies and feedback to evaluators are vital components of SSHAC Level 3 and 4 studies, but application of the tests briefly described in this Section should be viewed as optional and applied only where useful. In particular, as stated clearly in the Section 5.4, there is no possibility for testing that the center, the body and the range of technically defensible interpretations of available data, methods and models are represented in the final logic-tree.

The title of this Section refers to testing of hazard studies, and it is important to be clear with any test what its intended purpose is. In other words, it is important to understand fully what can be expected from its application. Although the terms 'verification' and 'validation' are widely (and loosely) used in this field, one should not expect to either verify or validate the hazard estimates, or indeed any of the component models that go into the hazard calculations. Oreskes *et al.* (1994) discuss verification and validation of numerical models in the Earth sciences and conclude that *"verification and validation of numerical models of natural systems is impossible. This is because natural systems are never closed and because model results are always non-unique."* Their arguments against the possibility of verifying models (i.e., proving that they are true representations of Nature) are made very clearly, starting with the fact that any attempt at verification will be based on data and *"what we call data are inference-laden signifiers of natural phenomena to which we have incomplete access. Many inferences and assumptions can be justified on the basis of experience (and sometimes uncertainties can be estimated), but the degree to which our assumptions hold in any new study can never be established a priori. The embedded assumptions thus render the system open."* Attempts at verification are further confounded by the problem of nonuniqueness, which is the fact that more than one model construction can produce a given output. Oreskes *et al.* (1994) go on to explain that *"A subset of the problem of nonuniqueness is that two or more errors in auxiliary hypotheses may cancel each other out. Whether our assumptions are reasonable is not the issue at stake. The issue is that often there is no way to know that this cancellation has occurred. A faulty model may appear to be correct. Hence, verification is only possible in closed systems in which all the components of the system are established independently and are known to be correct. In its application to models of natural systems, the term verification is highly misleading. It suggests a demonstration of a proof that is simply not accessible."*

Oreskes *et al.* (1994) then go on to discuss validation, which they note does not, in contrast to verification, necessarily denote establishment of truth. Their definition of validation is presented in these terms: *"a model that does not contain any known or detectable flaws and is internally consistent can be said to be valid."* They then issue the following warnings about interpreting validation of models as also validating their application: *"The term valid might be useful for assertions about a generic computer code but is clearly misleading if used to refer to actual model results in any particular realization. Model results may or may not be valid, depending on the quality and quantity of the input parameters and the accuracy of the auxiliary hypotheses."* Therefore, one can—and should—validate the computer codes used in PSHA by comparing results with established analytical solutions (*e.g.*, Thomas *et al.*, 2010); but this does not imply validation of hazard estimates obtained with these codes.

Oreskes *et al.* (1994) conclude that *"what typically passes for validation and verification is at best confirmation, with all the limitations that this term suggests."* They define confirmation as demonstration of agreement between observation and prediction, but note that *"confirmation is only possible to the extent that we have access to natural phenomena, but complete access is never possible, not in the present and certainly not in the future. If it were, it would obviate the need for modeling."*

The above statement patently applies to seismic hazard assessment, particularly for low annual exceedance frequencies, and from the outset it can be stated that verification and validation, in any strict sense, are not actually possible. Indeed, if they were then multi-expert assessments and logic-tree approaches would not be required because it would imply that input models are true representations of Nature and the true hazard level can be known with confidence. At most, tests can be conducted on components of the hazard model or on the hazard estimates themselves that could identify incompatibilities with observations. In other words, testing may invalidate components of the hazard model or even challenge the validity of parts of the hazard curve. There should be no expectation, however, that testing can provide conclusive validation of either.

Figure B-25 provides a simplified overview of the different options for testing in a PSHA study. The testing can either be applied to the SSC and GMC models that form the basic inputs to the hazard calculations, or to the hazard results themselves. Different tests will be appropriate for different output from the hazard calculations, depending on the annual frequency of exceedance (Anderson, 2010).

Figure B-25 indicates the data that can be used in the testing of models and also describes the questions that the tests can address. The uppermost boxes indicate tests that can be considered for the SSC and GM models in isolation, conducted prior to executing the hazard calculations. Testing the input models individually has advantages because tests on the hazard results may identify discrepancies but may not easily identify their source. Unfortunately, while testing of the SSC model is feasible and can be very effective, options for testing of the GM model are more limited, for reasons explained below.

The SSC model, in the very simplest terms, identifies the spatial and temporal distribution of future earthquakes of different magnitudes. The spatial aspect of the SSC model, provided all available data has been employed in its development, can only be decisively tested by observing future earthquakes over an infinite period to assess whether or not they occur at the expected rates and within the geographical limits of the different seismic sources. However, the general temporal and spatial patterns of seismicity implied by area source models can be tested using the method proposed by Musson (2004), in which the SSC model is used to generate synthetic earthquake catalogues. Statistical tests are then performed to determine whether these future projections are compatible with the historical earthquake catalog. The title of the paper by Musson (2004) refers to this as 'validation', whereas it would be more appropriately labeled as a test of internal consistency. Moreover, it is only useful if the hazard analyst has made the assumption of spatial stationarity for seismicity. Assumptions of temporal stationarity of seismicity may be difficult to check without a very long earthquake record. Schorlemmer et al. (2007) discuss an interesting experiment to test regional earthquake likelihood models (RELM) in California, but the test period is limited to 5 years. For the long-term estimates of seismic hazard appropriate to safety-critical facilities, the value of time-dependent seismic hazard models (which make use of such forecasts, at least in their temporal aspects) is not immediately evident.

A simple test of the internal consistency of an SSC model can be obtained by comparing the overall seismicity (in terms, for example, of the recurrence rate or the moment release rate) for the entire region, with that obtained by summing the contributions from all of the seismic sources and their associated recurrence relationships. If the two are significantly different, the SSC model may need to be revised or an explanation provided for why the future rate of seismicity is expected to be significantly different from the past rate.

The recurrence rates of earthquakes of different magnitudes are determined primarily from instrumental and historical earthquake catalogs on the one hand, and from geological data (paleoseismological investigations) on the other. These different data sources cover different periods of time with different degrees of reliability. The instrumental catalog is at best just over a century in length, and reliable for many parts of the world for only a few decades. The historical catalogue can extend back over hundreds or even thousands of years, but with inevitably higher levels of uncertainty the further one goes back in time. Moving even further back in time, geological data can cover thousands or even tens of thousands of years, but uncertainties inevitably arise with regard to identifying individual events and their time of occurrence. A useful exercise, wherever both types of data are available, is to compare the catalogs against geological data (e.g., Petersen et al., 2000), taking due account of the completeness of the catalog (e.g., Stein and Hanks, 1998).

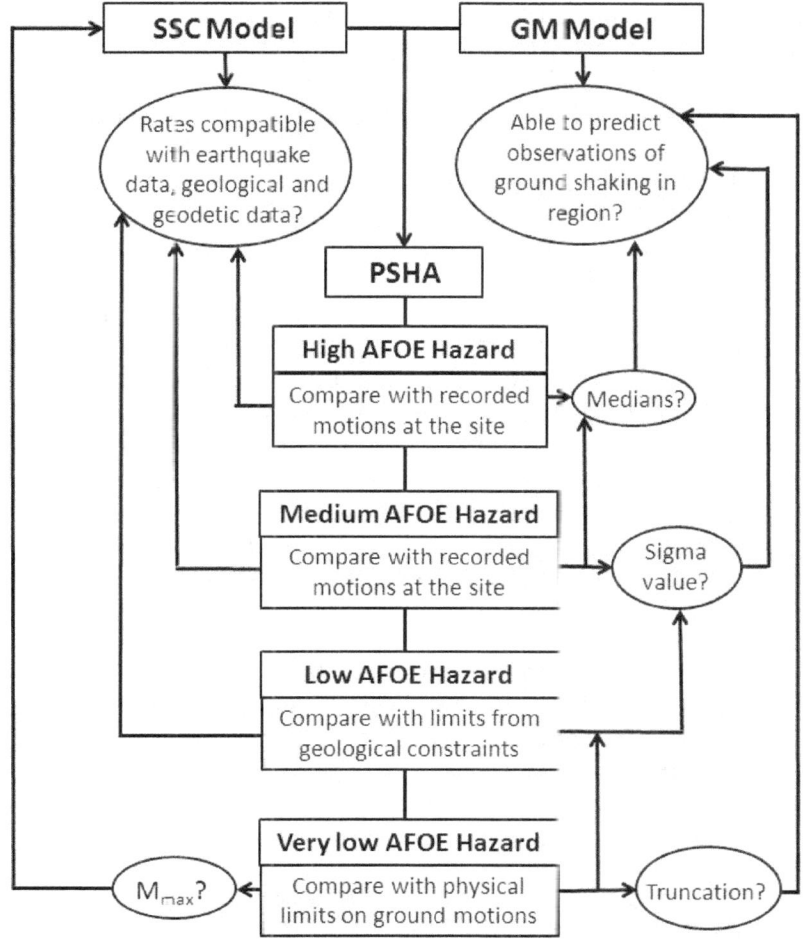

Note: AFOE is annual frequency of exceedance

Figure B-25.   Schematic Overview of Testing Options in PSHA.

Another data source that can be brought into the characterization of seismic sources is geodetic measurements from which crustal strain rates can be determined. The SSC model can also be explored in terms of consistency with deformations implied by instrumental and historical earthquake catalogs, geological deformation rates determined by paleoseismological investigations, and crustal deformation rates. In contrast with the seismological and geological data, geodetic data represent relatively very accurate measurements gathered over rather short periods of time, providing a snapshot of crustal processes. These rates can be compared with geological fault slip rates and the deformation implied by the total moment budget derived from the seismic source geometries and their associated recurrence relationships, bearing in mind that in some regions a large component of crustal deformation occurs aseismically. An example of such a comparison between seismologically- and geodetically-determined strain rates, made for a relatively large area, is presented by Pancha *et al.* (2006). Over smaller areas, the two types of data may be more difficult to reconcile. Such comparisons need to be aware of the very large uncertainties associated with the crustal strain rates determined by any of the approaches listed above, and as such these should probably not even be considered as 'tests'. The outcome of such comparisons must be interpreted with caution; and agreement between crustal strain rates estimated by two different types of data should not necessarily boost confidence in the SSC model.

Testing of ground-motion prediction equations is at least as difficult, given that their derivation invariably involves testing in terms of comparing the predicted distribution with the strong-motion recordings from which they are derived. In that sense, if the equation is being applied to the region for which it was derived, additional testing is unlikely to be feasible or useful, apart from inspecting the behavior of the model at the magnitude and distance limits to which it may be extrapolated in the hazard calculations. If the model is to be applied to another region, then its applicability can be tested if abundant local strong-motion recordings from the target region are available (e.g., Stafford *et al.*, 2008b). In such a case, the value of this exercise may be to justify using in models from other host regions if they have been derived from superior metadata and, by virtue of the host-region dataset or the modeling techniques, are possibly better suited to extrapolations to larger magnitudes and short distances. If there are relatively few data available in the target region, these can still be used to rank different ground-motion models in terms of their applicability (e.g., Scherbaum *et al.*, 2004b; Scherbaum *et al.*, 2009). If the local data are few, it is likely that they are also from smaller earthquakes, in which case interpretation of the results of such testing must take account of scaling issues. Intensity observations from the target region may allow the testing of the candidate ground-motion models for larger magnitude events (Delavaud *et al.*, 2009) although this is then subject to the considerable uncertainty in empirical relationships between macroseismic intensity and ground-motion parameters (e.g., Allen & Wald, 2009).

Once the hazard calculations have been performed, tests can be applied to the hazard curve, as indicated in Figure B-25. For short return periods, the hazard curve may be compared with a plot of ground-motion amplitude against frequency of occurrence constructed using multiple recordings of ground-motion at a single site (e.g., Ordaz and Reyes, 1999). Because the very first strong-motion accelerograms were recorded in 1933, the time span over which data are available anywhere for such an exercise is necessarily limited to at most a few decades.

Beauval *et al.* (2008) concluded that from a mathematical perspective, the period of time for which instrumental observations are required to estimate the true rate of ground-motion exceedances with a given degree of uncertainty is many times longer than the return period. To determine the ground motions with 475-year return period with a 20% uncertainty (coefficient of variation), for example, would require 12,000 years of recordings at the site. Discrepancies identified in such tests at a single site may point to problems with the recurrence relationship, although discrepancies could equally arise from the behavior of the seismic sources during the period of observation being unrepresentative of the long-term seismicity pattern. Interestingly, Stirling and Gerstenberger (2010) found that the agreement between the hazard curve and strong-motion recordings was improved when the former was calculated without aftershocks being removed from the earthquake catalog.

On the other hand, discrepancies may also reflect issues related to the median ground-motion predictions, and in particular the ability of the ground-motion model to predict local site effects. Site-specific effects can be partly removed from such tests by considering several sites simultaneously, as in the study by Fujiwara *et al.* (2009) who used 10 years of instrumental recordings from K-NET, transformed to JMA intensities, to 'validate' the national seismic hazard maps in Japan. Ward (1995) originally proposed an area-based approach, whereby if probabilistic seismic hazard is mapped over a region showing ground motions with X% probability of being exceeded in Y years, it can be tested by calculating the percentage of the total area where these accelerations are exceeded during Y years. Such multi-site tests, however, should take account of the fact that standard PSHA treats total sigma (both the inter- and intra-event components) entirely as earthquake-to-earthquake variability without separating out spatial variability. This simplification will lead to higher ground-motion amplitudes overall than would be expected in reality (Crowley and Bommer, 2006).

Intensity observations will often be available for much longer periods than those covered by instrumental observations of shaking, offering the possibility of testing the hazard curve at somewhat longer return periods. The basis of any such method is a comparison of the outcome of PSHA for a given return period with observed intensities over the same period, whether applied to a single site or over a region. This can be done by performing the PSHA in terms of macroseismic intensity (e.g., Albarello and D'Amico, 2005) but such hazard estimates are of very limited use for engineering applications. For any comparison with standard PSHA, conversions need to be made using relationships between macroseismic intensity and ground-motion parameters (e.g., Stirling and Petersen, 2006). These conversions introduce an additional degree of uncertainty, over and above that related to assumptions regarding completeness and accuracy of the historical intensity database. In the study of Miyazawa and Mori (2009) for Japan, the instrumental measurements of intensity were converted to JMA intensities, in the same way as in the study of Fujiwara et al. (2009).

To some extent, the use of intensity data to test PSHA results could be considered somewhat circular, because the intensity data is likely to have been used to develop the input to the hazard model. On the one hand, historical intensity data is used to extend the earthquake catalog back in time, inferring locations from the isoseismal pattern and estimating magnitudes from empirical relationships (e.g., Bakun and Wentworth, 1997). Additionally, as noted above, the intensity data may have been used to test the applicability of different ground-motion prediction equations. This circularity can be avoided by not using the intensity data to construct the original hazard model, but this is unwise because it would mean impoverishing the hazard model only so that it could be tested by data which when taken in isolation is subject to many uncertainties.

As indicated in Figure B-25, discrepancies identified in tests of PSHA based on comparisons with historical intensities may be related to the recurrence relationships developed or to the ground-motion prediction model used, both in terms of medians and to some degree the associated sigma value.

The hazard at longer return periods can only be tested against geologic data, such as precariously balanced rocks and other unstable features that could be toppled by intense ground shaking (e.g., Brune, 1999). By dating these features, and thus determining the time for which they have existed in a precarious state, and then determining the levels of shaking that would be expected to cause them to topple, limiting values on long-term hazard estimates may be estimated (e.g., Stirling and Anooshehpoor, 2006). If the hazard estimates exceed the toppling motions, then it may indicate that the recurrence rates of large earthquakes have been overestimated. However, the ground-motion models are generally interpreted to be the cause of such over-estimation. Rather than the median values being in error (although that cannot be discounted) most studies have concluded that the problem lies in the sigma values, which are overestimated due to the ergodic assumption (Anderson and Brune, 1999). As noted in Section B.3.2, recent studies have shown that sigma values for multiple earthquakes recorded at a single location are lower than those generally associated with GMPEs (e.g., Atkinson, 2006).

At very long return periods, the only constraints on hazard estimates are those that can be determined from fundamental physical models such as those based on the available stress levels in the ground and the maximum stress changes that can be transmitted through the crustal materials (Andrews et al., 2007). Limiting ground motions determined this way will generally not be related to any time period (other than that for which the geological conditions being modeled have been in existence) and hence they provide an absolute upper bound that the hazard estimates should not exceed regardless of the return period (e.g., Bommer et al., 2004). Should the ground motion estimates at very long return periods exceed these limits, the culprit is likely to be sought both in the sigma values used in the calculations and in the nature of the upper limit of the ground-motion residuals, which for current empirical ground-motion datasets are too small to characterize with any confidence (Strasser et al., 2008). Such a result may also raise doubts regarding the maximum earthquake magnitude, but the primary cause of the physical constraints being exceeded by the hazard estimates must lie in the value and nature of sigma.

When a PSHA has been completed, and any tests such as those outlined in the preceding paragraphs have been conducted, the analysts and others may be tempted to compare the new hazard results with any previous hazard estimates developed for the same location. While this is perfectly natural, it will often be the case that the new hazard estimates will be higher. This situation is often expressed erroneously in terms of the new hazard study having led to "higher hazard". However, there is only one true hazard curve for any location, and all that can be stated is that the new estimate of the hazard is higher than that found previously. More often than not, this will be because the hazard was underestimated in earlier studies (Bommer and Abrahamson, 2006). The most serious underestimations will arise from the hazard calculations being performed without integration across sigma; and correcting this omission will inevitably result in a major increase in the estimate of the hazard. More modest, but not insignificant, increases in hazard estimates will arise from early estimates of epistemic uncertainty being overly optimistic about the state of knowledge of earthquake processes. Increases in hazard estimates due to the introduction of sigma will be manifest in differences between median hazards. Once this correction is made, the increase in hazard estimates due to more effective capture of uncertainties can be seen by comparing mean hazard curves. The increase in hazard estimates due to more effective capture of epistemic uncertainty in some projects has led to unfounded statements about the SSHAC process leading to greater uncertainties. In reality, all it means in practice is that uncertainties were previously underestimated or ignored.

The discussion in the two preceding paragraphs concerns comparing new hazard estimates for a site with the results of previous hazard studies for the same location. In some cases, comparisons are made between a new hazard estimate at one site with an existing hazard assessment at another site in the same region. Comparisons of the hazard estimates at one location with those at another need to take account of the all of the above issues (regarding how the assessment at the other site was conducted), as well as differences in surface geology and the influence of any local seismogenic sources affecting either of the sites. In general, comparisons between nearby sites are problematic.

In some cases, comparison has been made between hazard estimates at widely separated locations. Comparisons with hazard estimates at remote locations, including in other countries, are generally made with preconceived ideas regarding the relative levels of seismicity (and hence hazard) between the two regions. Unless the hazard at the other location has been calculated using a correctly executed PSHA without unjustified truncations of sigma and with defensible estimates of epistemic uncertainties, such comparisons are very likely to be seriously misleading. Comparisons of hazard estimates can be useful if there is a clear attempt to identify the sources of the differences by examination of both estimates: there should not automatically be an onus on those responsible for higher hazard estimates to justify their results without there also being a rigorous examination of what factors may have contributed to the lower estimates either in an earlier study or an assessment for another location. In general, comparisons at widely separated locations are highly problematic and are to be avoided.

In closing, it is worth briefly mentioning that comparing probabilistic seismic hazard results with the output of DSHA is not infrequently invoked as a 'sanity check'. Here again, if such comparisons are made then it is important to explain the differences. Such comparisons can be made more informative by using the models developed for the PSHA to estimate the return period associated with the ground motions specified by the DSHA.

B.5   Probabilistic Analysis of Other Hazards

The framework for PSHA outlined in Section B.3 can be generalized such that it can be used to perform the probabilistic analysis of any hazard. The key element is the development of a model that describes the possible occurrence of the triggering event in space (location) and time (frequency of occurrence). If the event causes perturbations remote from the source, then an additional model is required to mathematically describe the radiation and propagation of these perturbations to the site. A model may also be required for any local or site-specific modifications to the perturbations at the point of interest. Aleatory variability in each of these components needs to be quantified and the hazard calculations must integrate over the distributions representing this variability when the calculations are used to determine the annual frequencies of exceedance of the consequence of interest at the site of interest. Alternative models and parameters for describing the occurrence of the initiating event, the propagation of the effects, and the local site-specific modifications to the effect constitute epistemic uncertainties and should be represented in a logic-tree.

Figure B-26 illustrates how the two basic components of a probabilistic hazard model (event generation and effect propagation) apply to PSHA and to the probabilistic analysis of tsunami hazard and surface fault rupture hazard as examples. In each case, it is necessary to begin by selecting the parameter for quantifying the effect at the site.

HAZARD (Analysis)	TRIGGERING EVENTS (Characterization)	MODEL for PROPAGATION to the SITE	MEASURE of EFFECT at the SITE
Ground Shaking (PSHA)	**Earthquakes** (SSC model for spatial and temporal distributions)	GMPE (GMC model)	PGA, PGV, spectral accelerations
Fault Rupture (PFDHA)	**Earthquakes** (magnitude, fault mechanism and slip)	Model for decay of displacement with distance	Surface displacement (sense and amplitude)
Tsunami (PTHA)	**Earthquakes** (location, size and frequency) **Landslides** (location, size, velocity and frequency)	Wave propagation and inundation	Sea level change (wave height or drawdown) Dynamic pressure Scour Debris impact

Figure B-26.   Generalized Framework of Probabilistic Hazard Analysis and Illustration of Application to Strong Ground-Motion, Surface Fault Rupture, and Tsunami.

Frameworks for the probabilistic analysis of surface displacements caused by fault rupture are presented by Youngs *et al.* (2003) and by Petersen *et al.* (2004). The paper by Youngs et al. (2003) draws a direct analogy between PSHA and probabilistic fault displacement hazard analysis (PFDHA). A PFDHA was conducted for the Yucca Mountain repository site in Nevada (Stepp *et al.*, 2001).

An approach to probabilistic tsunami hazard analysis (PTHA) was developed by PG&E (2010) based on a fully probabilistic implementation of the approach presented by Rikitake and Aida (1988). The latter study only considered offshore earthquakes as triggering events for tsunamis, whereas PG&E (2010) additionally included submarine landslides as tsunami triggers.

Probabilistic volcanic hazard analysis (PVHA) can also be conducted using the same general framework, in which the triggering events are volcanic dikes that would propagate into the repository (*e.g.*, Connor *et al.*, 2009). A PVHA was conducted as part of the characterization of the Yucca Mountain waste repository site in Nevada (SNL, 2008).

While a direct analogy can be drawn between probabilistic hazard analyses for tsunami (and for other fault-related hazards) with those for seismically-induced ground shaking, the analysis of the impact of secondary earthquake effects, such as liquefaction and earthquake-induced landslides, more closely resembles that of seismic risk analysis of structures and systems as illustrated in Figure B-1. Both liquefaction and seismically-induced slope instability are induced by ground shaking. Hence, the first step in the analysis of these hazards must be the assessment of the hazard in terms of ground shaking. The susceptibility of the soil deposits (for liquefaction) or the slope (for landslides) then needs to be characterized, which is analogous to the fragility functions derived for structures and components in probabilistic risk analysis. For liquefaction, the susceptibility is related to the degree of cohesion in the soil, the particle-size distribution, the relative density, and the degree of saturation, with loose, poorly-graded and saturated silts and sands being the most likely to liquefy during earthquake shaking. The susceptibility of natural and man-made slopes depends on factors such as the material strength, the slope angle, and the level of the ground water table, among many others.

Because there are common elements in the assessment of several correlated hazards, there are good arguments for conducting the analyses of these hazards (such as surface fault rupture and ground shaking, and possibly also liquefaction) together rather than as completely separate studies.

APPENDIX C: Perspective on this NUREG by Dr. Robert Budnitz, Chairman of the Senior Seismic Hazard Analysis Committee

This new report's stated purpose is to provide "practical implementation guidelines" to make the by now prominent "SSHAC" methodology more user-friendly and more effective. And the reason for this report – the need – is paradoxically because of both the remarkable success and the important shortcomings of the underlying SSHAC report itself. Let me explain.

When the US NRC, the US DOE, and EPRI came together to co-sponsor the Senior Seismic Hazard Analysis Committee (the "SSHAC Committee") in 1993, it was because there was a crisis of confidence in the field of probabilistic seismic hazard analysis (PSHA). This crisis had come about because two very prestigious, extensive, and very costly (multi-million-dollar) multi-expert PSHA studies, one sponsored by the US NRC and carried out under the leadership of experts at Lawrence Livermore National Laboratory, and the other sponsored by EPRI, had come to quite different overall conclusions about the seismic hazard facing the several dozen nuclear power plant sites in the eastern US. This difference occurred even though the two studies had involved many of the same seismic hazard experts, had used apparently similar technical and procedural methods to go about their work, and had been quite open about the "boundary conditions" or "rules" under which each of the two studies had been undertaken.

Subsequently, large numbers of both experts and non-experts in the PSHA field argued for a few years about why these differences had arisen. There were numerous hypotheses, more than one re-analysis, and extensive soul searching, but ultimately there emerged no resolution. Some "blamed" the expert elicitation process, some "blamed" the way the seismic experts had provided their interpretations of the data, or how these insights had been aggregated (or not), and some "blamed" an intrinsic bias on the part of one or the other (or both) of the sponsors, because the NRC's study produced "higher" seismic hazards than had the industry-sponsored study.

With this impasse facing everyone, the three sponsors (NRC, DOE, and EPRI) decided to try to reach resolution by putting together a panel of experienced scholars and practitioners in both seismic hazard and the use of experts, and asking them to try to resolve the issue. Originally, the idea seemed to be that the SSHAC committee would dig into the NRC/LLNL study and the EPRI study, and figure out which one had gone astray, and why –or perhaps to determine that both had gone astray. But as early as the first meeting, the SSHAC Committee itself realized that what was really needed was not a narrow resolution of the "EPRI vs. LLNL" problem, but rather a more fundamental examination of how to go about a large PSHA study, in terms of both its technical attributes and its procedural attributes, the latter coming down to how to make proper use of subject-matter experts.

The 7-member SSHAC Committee, supplemented by a comparable number of consultants and kibitzers who attended many of the meetings and made major intellectual contributions although they were not SSHAC members per se, then spent what amounted to about two and a half years of work (early 1993 to late 1995) figuring out how to go about a large and complex multi-expert PSHA study, and then how to explain it. (The final SSHAC report was published by the NRC in mid-1997, but was completed by late 1995, after which almost one and a half years went by prior to final publication awaiting a review by a specially appointed review panel of the National Research Council/National Academy of Sciences.) The SSHAC team's work was intense – meetings every month or two, extensive consultations in-between, several workshops attended by numerous others, some trials of the new methodology, lots of draft position papers some of which survived to form part of the ultimate report but some of which didn't, and quite a bit of back-and-forth about what we were trying to do and why. And it was all "out in the open": namely, the SSHAC deliberations and trial ideas were public at least to the large group of others (including the National Research Council panel) who were following our work and chiming in about it.

The SSHAC group early-on decided that a valid PSHA can be done at several different "levels" of detail and cost, after which we justifiably gave most of our attention to what we called the "Level 4" SSHAC process, somewhat less to the "Level 3" process, and almost no attention at all to the other two lower Levels.

Early on, the SSHAC group (not just the Committee but the consultants and kibitzers) all agreed that the major area where guidance was most needed was on the procedural side, not the technical side of how to go about a PSHA. After that realization sank in, the rest was at least a well-defined task, because we then concentrated on that topic intensely - not exclusively, mind you, because there is a lot of methodology guidance, especially in the SSHAC report's appendices, on technical topics in PSHA that get into gory detail about some of the seismic-source and seismic-ground-motion-attenuation aspects too. But the phrase "the rest was at least a well-defined task" in the preceding sentence does not capture the difficulty of the process we faced of writing down new and different guidance on procedures for how to use experts. Because we realized that the new SSHAC procedure was quite different from what had become known as the classical "expert elicitation process" in the literature, in that we emphasized interaction among the experts instead of assuring their full independence, the challenge of developing and defending the new approach was intensely and emotionally consuming. The group worked very well together, but there were definitely strong positions held and expressed by individuals whose stature made everyone else listen carefully. And nobody would have wanted it any other way.

In the end, what emerged was a methodology that we now call the "SSHAC procedure" for PSHA. We were and are (justifiably) proud of it, again not just the 7 SSHAC members but also the group of a half-dozen or so others who made major intellectual contributions to the deliberations. However, the one thing we all knew for sure was that, because our approach was new, controversial, and untested, it would definitely need a few trial runs before we'd all feel confident in the validity of the approach.

This background describes the need for the current report. By the time that more than a decade had passed, it had become obvious that clarification was badly needed for some issues that the SSHAC committee could not have known about or anticipated (they were too detailed), or in a few cases for some issues where the SSHAC guidance turned out to be unclear or ambiguous. And in a few areas there was a clear benefit to be had from developing guidance directly derived from the experience of the several SSHAC-type studies that were by then in the literature.

Hence this report. As I wrote above in the opening paragraph of this "Foreword", ".....the reason for this report – the need – is paradoxically because of both the remarkable success and the important shortcomings of the underlying SSHAC report itself." I am sure that the SSHAC committee and our consultants and kibitzers are all as proud of the original work as I am. I am also sure that all of us are grateful for this new work that helps make the methodology more accessible and useful.

And finally, I am sure that all of the other SSHAC members (George, Dave, Lloyd, Kevin, and Pete) are as grateful as I am to the brightest shining light in the group, the light that has been extinguished by the death of Allin Cornell. Nobody contributed more or is missed more. This new work, like the original SSHAC work, is a tribute as much to Allin as to all of the rest of us combined.

Robert J. Budnitz, Lawrence Berkeley National Laboratory
Chair, 1993-1997, Senior Seismic Hazard Analysis Committee

NRC FORM 335 (12-2010) NRCMD 3.7	U.S. NUCLEAR REGULATORY COMMISSION **BIBLIOGRAPHIC DATA SHEET** *(See instructions on the reverse)*	1. REPORT NUMBER (Assigned by NRC, Add Vol., Supp., Rev., and Addendum Numbers, if any.) NUREG 21 7, Rev 1

2. TITLE AND SUBTITLE	3. DATE REPORT PUBLISHED	
Practical Implementation Guidelines for SSHAC Level 3 and 4 Hazard Studies	MONTH April	YEAR 2012
	4. FIN OR GRANT NUMBER	

5. AUTHOR(S) Annie M. Kammerer Jon P. Ake	6. TYPE OF REPORT Final
	7. PERIOD COVERED *(inclusive Dates)*

8. PERFORMING ORGANIZATION - NAME AND ADDRESS (If NRC, provide Division, Office or Region, U. S. Nuclear Regulatory Commission, and mailing address; if contractor, provide name and mailing address.)

Division of Engineering,
Office of Nuclear Regulatory Research, U.S. Nuclear Regulatory Commission,
Washington, D.C. 20555-0001

9. SPONSORING ORGANIZATION - NAME AND ADDRESS (If NRC, type "Same as above", If contractor, provide NRC Division, Office or Region, U. S. Nuclear Regulatory Commission, and mailing address.)

Same as Above

10. SUPPLEMENTARY NOTES

11. ABSTRACT (200 words or less)

10 CFR 100.23, paragraphs (c) and (d) require that the geological, seismological, and engineering characteristics of a site and its environs be investigated in sufficient scope and detail to permit an adequate evaluation of the Safe Shutdown Earthquake (SSE) Ground Motion for the site. In addition, 10 CFR 100.23, paragraph (d)(1), "Determination of the Safe Shutdown Earthquake Ground Motion," requires that uncertainty inherent in estimates of the SSE be addressed through an appropriate analysis such as a probabilistic seismic hazard analysis (PSHA). In response to these requirements, in 1997, the U.S. Nuclear Regulatory Commission published NUREG/CR-6372, Recommendations for Probabilistic Seismic Hazard Analysis: Guidance on Uncertainty and the Use of Experts. Written by the Senior Seismic Hazard Analysis Committee (SSHAC), NUREG/CR-6372 provides guidance regarding the manner in which the uncertainties in PSHA should be addressed using expert judgment. In the 15 years since its publication, NUREG/CR-6372 has provided many PSHA studies with the framework and guidance that have come to be known simply as the "SSHAC Guidelines." The information in this NUREG is based on recent efforts to capture the lessons learned in the PSHA studies that have been undertaken using the SSHAC Guidelines. As a companion to NUREG/CR-6372, this NUREG provides additional practical implementation guidelines consistent with the framework and higher-level guidance of the SSHAC Guidelines.

12. KEY WORDS/DESCRIPTORS (List words or phrases that will assist researchers in locating the report.)	13. AVAILABILITY STATEMENT
Probabilistic Seismic Hazard Analysis PSHA SSHAC Senior Seismic Hazard Analysis Committee Uncertainty Expert Assessment	unlimited
	14. SECURITY CLASSIFICATION
	(This Page) unclassified
	(This Report) unclassified
	15. NUMBER OF PAGES
	16. PRICE

UNITED STATES
NUCLEAR REGULATORY COMMISSION
WASHINGTON, DC 20555-0001

OFFICIAL BUSINESS

NUREG-2117, Rev. 1

Practical Implementation Guidelines for SSHAC Level 3 and 4 Hazard Studies

April 2012